Learning Abstract Algebra with ISETL

Ed Dubinsky   Uri Leron

Learning Abstract Algebra with ISETL

DOS Diskette Provided

Springer-Verlag

New York Berlin Heidelberg London Paris
Tokyo Hong Kong Barcelona Budapest

Ed Dubinsky
Departments of Curriculum &
  Instruction and Mathematics
Purdue University
West Lafayette, IN 47907
USA

Uri Leron
Department of Science Education
Technion
Israel Institute of Technology
32000 Haifa
Israel

With 2 Illustrations

Mathematics Subject Classification (1991): 13-01, 20-01

Library of Congress Cataloging-in-Publication Data
Dubinsky, Ed.
    Learning abstract algebra with ISETL / Ed Dubinsky, Uri Leron.
      p. cm.
    Includes bibliographical references and index.
    ISBN 0-387-94104-5
      1. Algebra, Abstract—Computer-assisted instruction.   2. ISETL
  (Computer program language)   I. Leron, Uri.   II. Title.
  QA162.D83   1993
  512′.02′078—dc20   93-2609

Printed on acid-free paper.

Production managed by Natalie Johnson; manufacturing supervised by Vincent Scelta.
Camera-ready copy prepared using the authors' LaTeX files.
Printed and bound by R.R. Donnelley & Sons, Harrisonburg, VA.
Printed in the United States of America.

9 8 7 6 5 4 3 2 1

ISBN 0-387-94104-5 Springer-Verlag New York Berlin Heidelberg
ISBN 3-540-94104-5 Springer-Verlag Berlin Heidelberg New York

Contents

4 The Fundamental Homomorphism Theorem   119

Comments for the Student

Working to learn mathematics

This book is very different from other mathematical textbooks you have used in the past. Using any textbook represents an implicit "contract" between the writer and the reader, between the instructor and the student. In our "contract" we ask more from you and in turn we promise you will get more.

We ask you to work hard. And we really mean *work*. You will not be a passive recipient of mathematical knowledge. Rather, you'll be actively involved in *doing* things (mainly on the computer) and in discussing them with your fellow students. In fact, you and your colleagues will be, collectively, *constructing your communal and personal knowledge*. So when we say "work", we mean more than just putting in hours and doing the assignments. We are asking you to THINK — not only about the mathematics, but about what is going on in your mind as you try to learn Abstract Algebra. We are, in fact, asking you to take into your own hands the responsibility for your own learning and that of your colleagues.

And in return, we promise you that this book will make it possible for you to learn the mathematics meaningfully. If you use it properly and if your course is consistent with the principles on which this book is written, then you will begin to own your mathematical ideas, you will become an active learner, and you will learn a great deal of mathematics. You will see that abstract mathematical concepts start to make sense to you. You will be able to understand these concepts and even succeed in proving things about them. Less often will you find yourself stuck, staring at the symbols as if they were just so many ink stains on the paper. Indeed, we really can promise you that the symbols of mathematics will have meaning for you because, in working with the computer and reflecting together with your colleagues on that work, you will have constructed the meanings that the symbols represent. You will begin to control mathematics, rather than be controlled by it.

And perhaps we can promise you one more thing. If you are successful in this course, you will begin to see some of the beauty of mathematics. You will become initiated to a way of thinking that goes beyond utility and has for centuries fascinated some of the most powerful minds that the human species has produced. You will touch a little bit these minds and stand on some very tall shoulders. We think you will be very pleased with what you see.

Constructing mathematical ideas

In using this book to learn abstract algebra, much of what you will be doing is constructing mathematical ideas on a computer. You will write small pieces of code, or "programs" that get the computer to perform various mathematical operations. In getting the computer to work the mathematics, you will more or less automatically learn how the mathematics works! Anytime you construct something on a computer then, whether you know it or not, you are constructing something in your mind.

This is a fairly new approach to learning mathematics, but it has proved effective in a number of courses (including abstract algebra) over a period of several years. It works a lot better if you get into the spirit of things. In order to learn mathematics you have to somehow figure out most ideas for yourself. This means that you will often find yourself in situations where you are asked to work a problem without ever having been given an explanation of how to do it. We do this intentionally. The idea is to get you to think for yourself, to make the mathematical ideas your own by constructing them yourself, not by listening to someone talk about them.

A lot of people in this situation have a natural question. "If I am asked to solve a problem, or do something on the computer, and I am not told how to do it, what should I do?" The answer is very simple. Make some guesses (better yet, conjectures) and try them out on the computer. Ask yourself if it worked — or what part of your guess worked and what part did not. Try to explain why. Then refine your guess and try again. And again. Keep repeating this cycle until you understand what is going on. The most important thing for you to remember is not to think of these explorations in terms of success and failure. Whenever the computer result is different from what you expected, think of this as an opportunity for you to improve your understanding. Remember: instead of just being stuck, not knowing what to do next, you now have an opportunity to experiment, to make conjectures and try them out, and to gradually refine your conjectures until you are satisfied with your understanding of the topic at hand.

Another piece of advice is to *talk* about your work. Talk about your problems, your solutions, your guesses, your successes and your failures. Talk to your fellow students, your teachers, to anyone. It is not enough in mathematics to do something right. You have to know what you are doing and why it was right. Talking about what you are doing is a good way of trying to understand what you did. When we teach this course, our students always work in teams, whether they are doing computer activities or paper and pencil homework assignments. And they always talk about what they are doing.

Executing mathematical expressions on the computer and in your mind

Writing definitions and proofs and solving mathematical problems is like writing programs in a mathematical programming language and executing them in your head. Most people find this very hard to do. In this book we offer you a way to have the computer help in executing these mathematical expressions, so that you can always test your conjectures and compare your expectations with the results on the computer screen. Our experience shows that, as students become better at carrying out mathematical activities on the computer, their ability to "run" them in their mind improves as well.

Your computer work will be with the programming language **ISETL**. The nice thing about this language is that the way it works is very close to the way mathematics works. When you have written some **ISETL** code and you are running it, try to think about what the computer is doing and how it manipulates the objects you have given it. When you do this, you are actually figuring out how some piece of mathematics works. Another nice thing about this language is that learning to program in it is very similar to learning the mathematics involved. There's very little programming "overhead".

If you are not already familar with the computer system that you will be using, then you should expect to spend some hours practicing. If you are really new to this sort of thing, then, at first, it will seem very strange to you and things might go very slowly. Don't be discouraged. Everybody starts slowly in working with computers, but things get better very quickly.

Learning with this book

The book consists of six chapters and each chapter is divided into sections (usually three or four). The structure of individual sections in this book reflects our beliefs (supported by contemporary theory and research) on how people learn best. We believe (together with Dewey, Montessori, Papert, and Piaget) that people learn best by *doing* and by *thinking* about what they do. The abstract and the formal should be firmly grounded in *experience*.

Thus each section starts with a substantial list of *activities*, to be done in teams working on a computer. These are intended to create the experiential basis for the next learning stages. The activities in each section are followed by discussions, introducing the "official" subject matter. Some of these discussions read like a standard mathematics book with definitions, examples, theorems and proofs. But there are important differences. Explanations in the text are often only partial, raising more questions than they answer. Sometimes, an issue is left hanging and only you, the reader, can supply the missing link. Sometimes the discussion even uses ideas which will only be discussed officially some pages later. Once again, this style represents our realization that mathematical ideas can not be *given* to you. You must make them yourself. All we can do is try to create situations in which you are likely to construct appropriate mathematical ideas.

One reason why this kind of discussion works is because in reading the text you are not being introduced to totally unfamiliar material. Rather, it is just a more general and formal summary of what you have previously experienced and talked about in doing the activities. It summarizes and generalizes *your own experiences*. It is very important to remember that the activities are meant for *doing*, not just reading, and for doing *before* reading the discussions in the text (however, you will not be penalized if you decide to peek ahead). It is also very important to remember that the main benefit you get from the activities is due to the time and effort you have spent on them. It doesn't really matter so much whether you have actually found all the right answers. We repeat that the main role for the activities is to create an experiential basis, an intuitive familiarity with the mathematical ideas. The right answers will come after you have read the text, or after you have discussed matters in class and with your colleagues. In a few cases, however, there are some "right answers" that represent very deep ideas and you may not get them for a long time. Another part of learning mathematics is to learn to live with ideas that you only understand partially, or not at all. Through activities and discussion, you will come to understand more and more aspects of a topic. Eventually, it will all fit together and you will begin to understand the subject as a whole.

After the discussion in a section come the exercises. These are fairly standard because they come after you have had every opportunity to construct the mathematics in the section. The purpose of the exercises is to help you solidify your knowledge, to challenge your thinking, and to give you a chance to relate to some mathematical ideas that were not included in the section. You will find no cases in which the text lays out a mechanical procedure for solving a certain class of problems and then asks you to apply this procedure mindlessly to solve numerous problems of the same class. We omit this kind of material because, as must be clear to you by now, we believe that sort of interaction is of little help in real learning of mathematics.

Last words of wisdom

We have had our say, now its *your* turn to roll up your sleeves and start working. We hope that as you struggle through the activities, the text and the exercises, and especially as you struggle with the frustration which inevitably must accompany any meaningful learning of such deep material, you will not lose the Grand View of what you are doing—namely, learning successfuly and meaningfully one of the most beautiful albeit difficult pieces of mathematics. Especially we hope that you will succeed in maintaining an attitude of play, exploration and wonderment, which is what the spirit of true mathematics is all about.

Ed Dubinsky
Uri Leron

January, 1993

Comments for the Instructor

Teaching a course with this book

This book is intended to support a constructivist (in the epistemological, not mathematical sense) approach to teaching. That is, it can be used in an undergraduate abstract algebra course to help create an environment in which students construct, for themselves, mathematical concepts appropriate to understanding and solving problems in this area. Of course, the pedagogical ideas on which the book is based do not appear explicitly in the text, but rather are implicit in the structure and content.

In a sense this book lies somewhere in between a traditional text that supports a lecture method of teaching, and a book such as Halmos' *A Hilbert Space Problem Book* that can support a Moore-style approach. The ideas in our text are not presented in a completed, polished form, adhering to a strict logical sequence, but roughly and circularly, with the student responsible for eventually straightening things out. On the other hand, the student is given considerable help in making mathematical constructions to use in making sense out of the material. This help comes from a combination of computer activities, leading questions and a conversational style of writing. It should be noted that although it is assumed that each learning cycle begins with activities, the students are not expected to *discover* all the mathematics for themselves. In fact, since the main purpose of the activities is to establish an *experiential basis* for subsequent learning, anyone who spends a considerable time and effort working on them, will reap the benefits whether they have discovered the "right" answers or not.

It is also important to point out that the book is *not* primarily intended as a reference. Our main concern has been writing a book that will best facilitate a student's first introduction into abstract algebra. We see nothing wrong if, after learning the material through a course based on this book, a student uses some traditional text that is more suitable as a reference for someone who has already been through a first introduction.

In teaching abstract algebra courses based on this book, we are finding that our approach appears to be extremely effective for most students (omitting, perhaps, the top and bottom 5% ability group), bringing them much more into an understanding of the ideas in this subject than one would think possible from the usual experience with this course. For the superior student, the exercise set is strong enough to challenge and whet an appetite for more advanced mathematics. For all levels, we find that the students who go through our course develop a more positive attitude towards mathematical abstraction and mathematics in general.

Finally, before describing the structure of the book, we should point out that the text is only part of the course. We have available a package to aid instructors in using our approach. This package includes: a disk with the required software (running on Macintosh or PC); documentation for the software; complete sets of assignments, class lesson plans, and sample exams; answer keys; and information on dividing students into teams. We have also been experimenting with an alternative method for introducing students to **ISETL** more quickly. This method, which may be suitable for some classes, utilizes four worksheets and discussions based on them, as a substitute for the more extensive Chapter 1. These worksheets are also included in the package for instructors. Please contact the authors for more information.

The ACE cycle

The text is divided into sections, each intended to be covered by an average class in about a week. Each section consists of a set of activities, class discussion material, and a set of exercises.

Activities. These are tasks that present problems which require students to write computer code in **ISETL** representing mathematical constructs that can be used to solve the problems. Often, an activity will require use of mathematics not yet covered in the text. The student is expected to discover the mathematics or even just make guesses, possibly reading ahead in the text for clues or explanations.

Class Discussion Material. These portions contain some explanations, some completed mathematics and many questions, all taking place under the assumption that the student has already spent considerable time and effort on the activities related to the same topics. Our experience indicates that with this background, students can relate much more meaningfully to the formal definitions and theorems. Each unanswered question in the text is either answered later in the book or repeated as an explicit problem in the exercises. Our way of using this discussion material in a course is to have the students in a class working together in teams to solve paper and pencil problems, mainly suggested by the open questions in the text. This largely replaces lectures which occur only as summaries after the students have had a chance, through the activities and discussions, to understand the material.

Exercises. These are relatively traditional and are used to reinforce the ideas that the students have constructed up to this point. They occasionally introduce preliminary versions of topics that will be considered later.

Covering the course material

Though the teaching method supported by this book is novel, the selection of material is standard. The book contains the material on groups, rings and fields usually covered in a one-semester course, though we would be happier if we could stretch it over 1.5 or 2 semesters. We feel that for many students, going beyond the material on group theory in one semester interferes with their ability to advance beyond a superficial understanding of Abstract Algebra.

The first chapter covers all the necessary knowledge and practice on **ISETL**, while at the same time introducing some of the mathematical systems (such as modular arithmetic and permutations) that will be used most often in the remainder of the book. Chapter 2 introduces the group concept, while Chapters 3 and 4 take the student deeper into group theory, notably Lagrange's theorem and the fundamental theorem of homomorphisms. Chapter 5 introduces ring theory, always building on the analogy between groups and rings. Finally, Chapter 6 brings the theory of factorization in integral domains, presented as a generalization and abstraction of the well-known facts about the integers and about polynomials. It culminates in the fundamental theorem of field theory about the possibility of adjoining roots of polynomials over fields.

Ed Dubinsky
Uri Leron

January, 1993

Acknowledgments

The authors are grateful to the following people who read earlier versions of the book and made valuable comments:

David Chillag, Orit Hazzan, Robert Smith, and Alfred Tang.

We would also like to acknowledge the contributions to our work which have been made by the members of our Abstract Algebra Project: Jennie Dautermann, James Kaput, Robert Smith, and Rina Zazkis. This project has been supported by a grant from the National Science Foundation and we would like to express our gratitude for that.

Finally, we would like to thank the many students who have been willing to subject themselves to the very unusual kind of undergraduate mathematical experience supported by this text. We hope that their growth in understanding of the beautiful subject of abstract algebra has justified their hard work and their courage to plunge into the unknown.

1
Mathematical Constructions in ISETL

1.1 Using ISETL

1.1.1 ACTIVITIES

1. Use the documentation provided for your computer to make sure that you can answer the following questions.

   (a) How do you turn the thing on?

   (b) How do you turn the thing off?

   (c) How do you enter information? From the keyboard? A mouse? Disks?

   (d) How do you move around the screen? With keys? A mouse?

   (e) How do you make files and how are they organized?

   (f) How do you save files or discard them?

   (g) How do you make back-ups?

2. Use the documentation provided for **ISETL** to make sure that you can answer the following questions.

   (a) How do you start an **ISETL** session?

   (b) How do you end an **ISETL** session?

   (c) How do you enter information to the system by typing directly?

(d) How do you enter information to the system by working from a file and transfering?

(e) How do you change things, correct errors, add or delete material?

(f) How do you save the work that you do in a session?

(g) How do you print from files or windows?

3. Following is data that can appear on the screen when you are in **ISETL**. A line beginning with a > or >> prompt must be entered by you in response to the prompt. The end of such a line indicates that you must press **Return** or **Enter**. The other lines are put on the screen by **ISETL**. Start **ISETL** and operate interactively to enter the appropriate lines and obtain the indicated responses.

```
>     7 +18;
25;

>      13*(-233.8);
-3039.400;
>     6 = 2 * 3;
true;

>     5 >= 2* 3;
false;
>     170
>>    + 237 - 460
>>    *2
>>    ;
-513;

>     n := 37 mod 23;
>     n;
14;
>     N;
OM;
>     p := -4 mod 23;
>     p;
19;
>     (n + p) mod 23 = 10;
true;
```

```
>     is_number(3.7);   is_number(3 < 4); is_boolean(3 < 4);
true;
false;
true;

>     A := "Abstract Algebra";
>     A(1);
"A";
>     A(4); A(9);
"t";
" ";
>     A(11); A(6); A(10);
"l";
"a";
"A";
>     is_string(A);   is_string(A(6));
true;
true;

>     A = A(10);
false;
>     B := "ABSTRACT"; C := "AB" + "STRACT";
>     B = C;
true;

>     "B" in "ABS"; "b" in "ABS";
true;
false;
```

4. Following is a list of items for you to enter into **ISETL**. Before entering, guess and write down what **ISETL**'s response will be. In case the response is different from what you predicted, try to understand why.

It may be more convenient for you to work with a file and copy items to the screen as you enter them. The specifics for doing this will vary with the system.

```
>    b := 10;
>    b;
>    b + 20;
>    b := b - 4; b; B;
>    is_defined(b); is_defined(B);

>    MaxEquals := (max(2,3)=2);
>    maxequals;
>    MaxEquals;
>    MaxEquals or (0.009758 < 0.013432);

>    (2/=3) and ((5.2/3.1) > 0.9);
>    (3 <= 3) impl (3 = 2 +1);
>    (3 <= 3) impl (not (3 = 2 +1));
>    (3 > 3) impl (3 = 2 +1);
>    (3 > 3) impl (not (3 = 2 +1));

>    7 mod 4; 11 mod 4; -1 mod 4;
>    (23 + 17) mod 3;

>    a := 0; b := 1; c := 2; d := 3;
>    a := d; b := c; c := b; d := a;
>    a; b; c; d;

>    is_integer(5); is_integer(-13);
>    is_integer(6/4); is_integer(6/3);
>    is_integer(3.6); is_integer(6 - 3);
>    is_integer(a); is_integer(a mod c);
>    is_boolean(true); is_boolean("true");
>    is_boolean(false); is_boolean("false");
>    is_boolean(2 < 3); is_boolean(-2 mod 3 > 0);
```

5. Following are some groups of **ISETL** code. Read each group, predict what its result will be and write down an explanation, in words, of what the code does. Then enter it into **ISETL** and compare what you get with your prediction. Explain any discrepancies.

   (a)
```
>    x := 4; y := 2;
>    if (x + y) mod 6 = 0 then
     ans := "Additive
     Inverses!"; end;
>    ans;
```

(b)

```
>      if (x + y) mod 6 = 0 then
>>          ans := "additive";
>>          ans := ans + "inverses!";
>>     end;
>      ans;
```

(c)

```
>      if (x * y) mod 6 = 1 then
>>              ans := "multiplicative inverses!";
>> else       ans := "not multiplicative inverses!";
>>     end;
>      ans;
```

(d)

```
>      if (x * y) mod 7 = 0 then
>>              ans := "not multiplicative inverses!";
>> elseif (x * y) mod 7 = 1 then
>>              ans := "Now this time they are!";
>>     end;
>      ans;
```

(e)

```
>      for i in [-9..9] do
>          writeln i, i mod 10;
>      end;
```

1.1.2 Getting started

Look over the documentation provided by your instructor and read through
the material in *Using ISETL 3.0: A Language for Learning Mathematics*.
Don't be afraid to ask questions — of anybody, whether they seem to know
what they are doing or not. Try anything you can think of and don't be
afraid to make mistakes. The only way to learn anything is to figure out
the reason for your mistakes.

Activities 1 and 2 list all of the things you need to do in order to run
ISETL on your computer system. The way **ISETL** works is that you enter
something (that means type and press the **Return** key) and the computer
responds. It makes two kinds of responses. One is internal and you don't
see anything on the screen. The other kind of response is to print something
on the screen. Depending on what you gave it, **ISETL** will respond with
a value that is the result of an operation your data caused it to perform,
some sort of error messsage, or just a > prompt which is a signal that some
internal activity is completed and **ISETL** is ready for more input.

Each kind of input has a structure with syntax rules to be followed before
the input is recognized by **ISETL** as a complete unit. It must be complete
before **ISETL** can work on it. Very often your input must be completed

with a semicolon (;). There are a few exceptions, but rather than list them here, we think it is better for you to see them in situations as you go along.

If the response that you get is a >> prompt, it means that **ISETL** did not recognize your input as complete and you must type something more to complete it. (Very often, especially in the beginning, this means that you have forgotten a semicolon (;) or an end;).

Here is a suggestion. Some people prefer to work from a file. That is, don't enter your code directly onto the screen in **ISETL**. On some systems, such as a Macintosh, it is easy to enter your code in a file and transfer it to **ISETL**. (Check the documentation to see how it is done.) This way you are maintaining a permanent record of what you have done. You should also save this file periodically. Then, if there is some power failure or other "glitch", you can use the contents of your file to restart without any serious loss of the work you were doing when the trouble occured.

Activities 3, 4, and 5 illustrate many of the **ISETL** constructions you will be using. Read the next two paragraphs for some explanations.

1.1.3 Simple objects and operations on them

Look at the code in Activities 3 and 4. You should understand the role of the two kinds of prompts, > and >>. If you press **Return** too soon, before your input can be recognized by **ISETL** as a complete unit, you just get a >> prompt that tells you to continue typing and enter the rest of your data. In most cases, you can continue on the same line without going back.

You will see integers in the code given in these two activities, and the standard arithmetic operations of addition (+), subtraction (-), multiplication (*) and division (/). You will also see numbers like -233.8 and -3.03940e+003 which are *decimal* and *floating point* representations of rational numbers. Most of our numerical work in this book will be with integers.

Notice that some, but not all, of the arithmetic operation on integers give an integer answer. The *type testers* at the end of Activity 4 can be used to tell what kind of value you are working with. Do you see why we say that the integers are *closed* with respect to addition, subtraction, and multiplication, but not division? What would we say about closure if we replaced "integers" by "positive integers"? Non-negative integers? What about rational numbers? (A number is called *rational* if it can be written as a/b where a and b are integers and $b \neq 0$.)

The **ISETL** values true and false are called *boolean* values. Roughly speaking, any operation that asks a yes-or-no question returns a boolean value.

An input such as

>   n := 37 mod 23;

does not cause anything to be printed on the screen except a new > prompt asking for more input. This is because the response is completely internal. **ISETL** has an internal representation of the variable n and this input causes it to set the value of that variable to be whatever is the result of 37 mod 23. We refer to this by saying that the value on the right hand side of := is *assigned* to the variable n. It is only after doing that, if you give the input

>   n;

that a response appears on the screen. **ISETL** interprets this input as a command to evaluate n, that is, to print on the screen the quantity 14 which is the "answer" to 37 mod 23.

This operation mod on integers is very important in Abstract Algebra. The expression a mod b, where a and b are integers, refers to the (positive) remainder that you get when you divide a by b.

A character string in **ISETL** is indicated by enclosing it in quotes " ". You can use function notation such as A(4) to access a particular character in a string. In the example given in Activity 3 what would be the value of A(9)?

There are other operations such as concatenating two strings (+) and testing for a character being in a string.

ISETL is case sensitive and will distinguish between a lower case and a capital letter. In some systems, however, this feature is ignored and you can use the examples in Activity 4 to see how it goes on your system.

The operation a impl b is very important. In mathematics it is written $a \Rightarrow b$. In any case, a and b must be expressions whose values are boolean (**true** or **false**.) How many possibilities then are there for the pair of values of a, b? What is the value of $a \Rightarrow b$ in each of these cases? Test your answer on the computer. Can you explain the various answers?

Were you surprised by the result of

```
>     a := 0; b := 1; c := 2; d := 3;
>     a := d; b := c; c := b; d := a;
>     a; b; c; d;
```

Do you think that this result is what the programmer had in mind? Probably not. Can you guess what was intended? How would you make it right?

1.1.4 Control statements

Activity 5 illustrates two of the most important control statements that you will be using in Abstract Algebra: the conditional if...then... statement and the iteration for loop. We expect that you have more or less figured them out for yourself in the activities. Here is a more formal explanation.

An if...then... statement begins with the key word if followed by an expression which has a boolean value followed by the key word then, followed by a list of commands. After that there can be any number (including none) of elseif phrases, consisting of the key word elseif, followed by a boolean valued expression, followed by a list of commands. Then there may or may not be a single else phrase , which consists of the key word else followed by a list of commands. The statement is completed by the key word end.

An if...then... statement evaluates each of its boolean expressions in logical order. If an expression evaluates to false, the commands under it are ignored. Once it encounters a boolean expression that has the value true, it does not look at any more phrases. It just performs the list of commands under it and that is the end of its response.

A for loop begins with the key word for followed by an *iterator*. An iterator is a domain specification of one or more variables in a tuple or set. You will see how tuples and sets work in the next section. For the example in Activity 5, the expression [-9..9] is the tuple which represents the sequence of integers from -9 to 9. After the iterator, the for loop has the key word do followed by a list of commands. The code for this structure is completed with the key word end.

A for loop performs each of the commands in its list, once for each value of the variable in the iterator.

1.1.5 EXERCISES

1. Write out explanations, in your own words, for each the following terms. Note that anything which is in typeset font is considered to be an **ISETL** keyword.

   (a) prompt

   (b) true

   (c) om

   (d) is_whatever

   (e) string

   (f) mod, div

   (g) boolean

   (h) if statement

   (i) for loop

   (j) input

   (k) objects, operations

(l) ;

(m) impl

2. Read the following code and follow the instructions and/or answer the questions listed after the code.

```
rp := om;
x := 12;
y := 18;
if is_integer(x) and is_integer(y) and x > 0 and y > 0
    then rp := true;
    for i in [2..(x .min y)] do
        if (x mod i = 0) and (y mod i = 0) then
            rp := false;
        end;
    end;
end;
write x, y, rp;
```

(a) Run the code several times, with different initial values for x,y.

(b) Write out an explanation, in your own words, of what this code does. In particular, explain how the code gets data to work on, what it does with the data, and what is the meaning of the result.

(c) Place this code in an external file. Exit **ISETL** and then re-enter **ISETL** to run this code without retyping it.

(d) What does it mean to say that this code tests its input?

(e) Suppose that you run this code and that the value of y you enter is always twice the value of x. Can you be sure of what the value of **rp** will be? Why?

(f) Add a statement to the code that will display a meaningful announcement about the result.

(g) List some relationships between the values of x and y for which you can always be sure of the value of **rp** at the end.

(h) Suppose that values a,b for x,y result in **rp** having the value **true** and this is still the case for values b,c for x,y. What will happen if you give x,y the values a,c?

3. Look at each of the following sets of **ISETL** code. Predict what will be the result if it is entered. Then enter it and note if you were right or wrong. In either case, explain why.

```
!rational off
2 + 3 * 2**3 - 10; 10 + 6/4;
2**3**2; (2**3)**2; 2**(3**2);
4/2; 4/3;
!rational on
10 + 6/4; (10 + 6)/4; 6/4 + 10; (6/4) + 10;
(4/2) + (4/3); (2/3) - (5/12); (1/4)*(3/5); (1/4)/(3/5);
!rational off
```

```
12 div 4; 12 div 5; 12 div -5; -12 div 5; -12 div -4; 12
div 0; 12 mod 4; 9 mod 4; 6 mod 4; 3 mod 4; -2 mod 4; -4
mod 4; -7 mod 4; 2 = 3; (4+5) /= -123; (12 mod 4) >= (12
div 4);
```

```
even(2**14); odd(187965*45);
max(-27,27); min(-27, max(27,-27));
abs(min(-10, 12) - max(-10, 12));
max(sgn(34)*(124 div -11)-2**5 + min(20, abs(-20)), -(31
mod 32));
```

4. Look at each of the following sets of **ISETL** code. Predict what will be the result if it is entered. Then enter it and note if you were right or wrong. In either case, explain why.

   (a) n := 58; (n div 3)*3 + n mod 3;

   (b) is_integer(-1020.0) and is_integer(-1020);

   (c) is_integer(9) impl is_string(234);

   (d) is_integer(-17) impl is_string("234");

   (e) is_string(234) or not(is_integer(9));

   (f) is_string("234") or not(is_integer(-17));

   (g) true impl true; true impl false; false impl true; false impl false;

5. Find the value of x mod 6 for x = -7, -6, ... , 6, 7.

6. Describe all possible integer values of a for which a mod 6 = 0.

7. Describe all possible integer values of b for which b mod 6 = 4.

8. Re-express each of the following in as simple a way as you can, using x, y, max, min and abs, assuming that the values of x and y are integers.

(a) `max(x,x);`

(b) `min(y,y);`

(c) `max(y, min(x,y));`

(d) `min(max(x,y), x);`

(e) `max(y, max(x,y));`

(f) `min(min(x,y), x);`

9. Write **ISETL** code that will run through all of the integers from 1 to 50 and, each time the integer is even, will print out its square.

10. Change your code in the previous problem so that instead of even integers, it will print out the square each time the integer gives a remainder of 3 when divided by 7.

11. Use **ISETL** to determine the larger of the fractions $\frac{2}{3} + \frac{8}{9}$, $\frac{4}{5} + \frac{6}{7}$. Do you see a pattern in the choice of the four fractions? Run several variations of the pattern and see if you can find a general rule.

1.2 Compound objects and operations on them

1.2.1 ACTIVITIES

1. Following is a list of items for you to enter into **ISETL**. Before entering, guess and write down what **ISETL**'s response will be. In case the response is different from your prediction, try to understand why.

```
T1 := [0..19]; T1;
T2 := [0,2..19]; T2;
T3 := [0,6..19]; T3;
T1(5); T2(5); T3(5);
#T1; #T2; #T3;

T4 := [1,2,"Ed","Uri",T2,3<2,["algebra","ISETL",
"textbook"]]; #T4;
T4(7);    T4(7)(2);
2 in T4;    false in T4;    "ISETL" in T4;

%+[1..10];    %*[1..6];    %or[2=1,2=2,2=3,2=4];

Z12 := {0..11}; Z12;
threeZ12 := {0,3..11}; threeZ12;
threeZ12' := {9,0,3,6,3}; threeZ12' = threeZ12;
```

```
!setrandom on
T1; T1; T1; T1;
Z12; Z12; Z12; Z12;

!setrandom off
T1; T1; T1; T1;
Z12; Z12; Z12; Z12;
```

2. Write a few paragraphs describing your experience with Activity 1. What did you predict? What happened? How do you explain what **ISETL** did? Rather than just reporting events in chronological order, try to organize your description in some logical order and suggest generalizations. Include a description of the main differences between using [..] (which constructs a tuple or sequence) and {..} (which constructs a set.) What is the effect of the directive !setrandom on/off?

3. Write out a verbal explanation of the result of giving each of the following input lines to **ISETL**.

```
Z20 := {a mod 20 : a in [-30..50]};
H := {g : g in Z20 | even(g)};
K := {(5*g) mod 20 : g in Z20 };
L := {g*h : g,h in Z20 | even(g) and h < 10};
HK := {(h * k) mod 20 : h in H, k in K};
#(Z20); #(H); #(K); #(HK);

p := [3,1,2]; q := [3,2,1]; r := [p(q(i)) : i in [1..3]];
r; S3 := {[a,b,c] : a,b,c in [1..3] | #{a,b,c}=3};
```

(The elements of S3 are called permutations and the tuple r formed from p and q is called their composition.)

```
H union K; H union HK; K union HK;
K inter H;    H inter HK;
H subset K;    HK subset H;    K subset HK;
H subset K; H subset HK; K subset HK;
Z20 - {0}; 0 in Z20; 0 in Z20 - {0};
S := pow({0,1,2,3}); {0,1} in S; {} in S;
arb(Z20); arb(Z20); arb(Z20); arb(Z20);
```

4. Write out a verbal explanation of the result of giving each of the following input lines to **ISETL**.

```
Z20 := {0..19};
S3 := {[a,b,c] : a,b,c in [1..3] | #{a,b,c}=3};

forall x in Z20 |(x + 0) mod 20 = x;
forall x in Z20 |(x + 3) mod 20 = x;

exists p in S3 | p(1) < p(2);
exists p in S3 | p(1) = p(2);
exists e in Z20 |(forall g in Z20 | (e + g) mod 20 = g);

forall g in Z20 |(exists g' in Z20 | (g + g') mod 20 =
0); forall p,q in S3 | [p(q(i)) : i in [1..3]] in S3;

choose e in Z20 | (forall g in Z20 | (e + g) mod 20 = g);
e := choose x in Z20 | (forall g in Z20 | (x + g) mod 20
= g); e;
```

5. Look at the following **ISETL** code and try to predict what the results would be. Then run the code and check. Explain any discrepancies between your predictions and the actual answers.

```
S1 := {2,"a",[1,2]}; S2 := {"a", 2,[1,2]};
S2' := {"a",[1,2], 2, "a"};
S1 = S2; S1 = S2'; S2 = S2';
T1 := [2,"a",[1,2]]; T2 := ["a", 2,[1,2]];
T2' = ["a",[1,2], 2, "a"];
T1 = T2; T1 = T2'; T2 = T2';
S := {0,1,{0,1}}; T := [0,1,{0,1}];
S = T;
```

6. Write **ISETL** code that will construct the following sets. Run your code to check that it is correct.

   (a) The set of all integers from 1 to 1000 whose squares mod 20 are greater than 14.

   (b) The set S4 of all **tuples** which represent permutations of the integers 1,2,3,4.

   (c) The set of all compositions of the **tuple** p with the **tuple** q where p,q run through all elements of S3.

   (d) The set of all elements of the form [[x,y],(x+y) mod 6] where x,y run through all the elements of Z6.

   (e) The set of all elements of the form [[p,q],r] where p,q run through all elements of S3 and r is the composition of p with q.

7. Write **ISETL** code that will test the truth or falsity of the following statements. Run your code to check that it is correct.

   (a) Every element of Z20 is even.

   (b) Every element of S3 is a `tuple`.

   (c) Some element of Z20 is a `tuple`.

   (d) Some elements of Z20 are odd.

   (e) Some component of every element of S3 is odd.

   (f) The product **mod** 20 of every pair of elements of Z20 - {0} is again in Z20 - {0}.

   (g) Every element of Z20 has a corresponding element which, when added to it **mod** 20 gives the result 0.

   (h) There is an element of Z20 which, when added to any element of Z20 does not change it.

8. Write **ISETL** code that will express the negation of the statements in Activity 7, parts (e), (f), (g), (h). One way to do this is to apply the operation **not** to the statement. Find another way.

   In general, what is the meaning of the "negation of a statement"?

1.2.2 Tuples

The **ISETL** object called a `tuple` is used to represent a finite sequence. The simplest way of doing this is a list of consecutive integers such as

$$[1,2,3,4]$$

or

$$[0..19] ;$$

or

$$[-30..50] ;$$

which represent, respectively, the sequences of integers from 1 to 4, 0 to 19 and −30 to 50.

It is also possible to construct a general arithmetic progression of integers by giving the first two and the last as in,

$$[0,2..19] ;$$

This represents the even integers between 0 and 19. The way this works is that on receiving such an input, **ISETL** will construct a sequence by starting with the given first and second terms. Subsequent terms are obtained by adding the constant difference which is obtained by subtracting the first term from the second. Terms are added as long as they do not exceed the given last term.

Can you imagine how to construct a decreasing sequence in **ISETL**?

The components of a `tuple` do not have to be integers. They can be any **ISETL** objects, including other `tuples`. Here are some examples. Run them in **ISETL** and make sure you understand what is going on.

```
[3, "three", 2+1];
x := 4; [x, 4, v, #([2,om,5]) = 3, [2,om,5]];
```

One of the most important facts about `tuples` is that their elements come in a fixed, definite order. Each time you evaluate a `tuple`, you get the same sequence in the same order. As a result of this, it is possible to access specified components of a `tuple`. For example, you saw in Activity 1 that the value of the expression T1(5) is the value of the 5th component of the `tuple` T1. What do you think would be the value of the expression T1(8)?

1.2.3 Sets

The **ISETL** object called a `set` is exactly the same as a finite set in mathematics. As you saw in Activity 1 it is possible to construct a `set` consisting of a sequence of consecutive integers and, more generally, any arithmetic progression.

Two important ways in which sets differ from tuples are that in sets the order of the elements, as well as how many times they are repeated, do not matter. Thus you saw in Activity 1 that the sets $\{0,3,6,9\}$ and $\{9,0,3,6,3\}$ are equal. Activity 5 is another illustration. In mathematics these two properties of sets are expressed concisely in the definition of when two sets are to be considered equal, that is, when they have precisely the same elements. More explicitly, sets A and B are equal if x in A implies x in B and vice versa. Can you see how the two properties above come out of this definition?

In ISETL there are two options for displaying sets on the computer screen. In the !setrandom on option (the one selected automatically when ISETL is started), each time a set is evaluated its elements are displayed in a different order. This option helps emphasize that a set remains the same even when its elements are "scrambled". In working in abstract algebra, however, it is more convenient in some situations if sets are always displayed in the same "natural" order, since it is then easier to recognize visually when two sets are the same. This option is chosen by entering !setrandom off. Of course these two options do not affect any of the properties of the sets, only how they are displayed on the screen.

1.2.4 Set and tuple formers

Our ability to construct examples of sets and tuples in **ISETL** can be enriched by using the *former*. Examples of these are given in the first group of lines in Activity 3. The structure is the same for both sets and tuples. For example, the set former {g : g in Z20 | even(g)} of Activity 3 evaluates to the set of all even elements in Z20. Similarly, the set former

$$\{g*h \ : \ g,h \ \text{in} \ Z20 \ | \ \text{even}(g) \ \text{and} \ h < 10\}$$

evaluates to the set of all products of an even element of Z20 by an element of Z20 which is smaller than 10.

In general the former has three parts. The first part is an expression. Every variable that appears in this expression must either have been previously given values or must appear in the second part of the former. The first part is completed with a colon (:) or a bar (|.) The second part is called the *domain specification*. This is a list of phrases of the form x in S, or x,y in S separated by commas. Here, x, y are variables, in is a key word, and S is an expression whose value is a set or tuple.

It is not necessary for every variable that appears in the list of domain specifiers to appear in the expression in the first part. It is, however, required that every variable appearing in the first part, which has not been previously given a value, must appear in the domain specification.

The last part of a former is optional. If present, it begins with the symbol | (or :) and is followed by a boolean expression, i.e., one whose value is true or false.

Here is a description of how **ISETL** constructs a set (or tuple) when it is given a former. **ISETL** iterates through all possible combinations of values of the variables defined in the domain specification. For each particular combination of values of the variables, the boolean expression in the third part is evaluated, if it is present. If it is not present, then the value is taken to be true. If the result of evaluating the boolean expression is false then nothing more is done and **ISETL** moves on to the next value of the variable(s). If the result is true, then **ISETL** evaluates the expression in the first part of the former and the result is included as an element of the set or component of the tuple being constructed. Thus, the value of a set former expression is the set of all values of the expression (obtained by iterating the variables through their domains) for which the condition holds.

In Activity 6 you have an opportunity to write some set formers.

We repeat in closing that a tuple can be used to represent a (finite) sequence. A sequence can be interpreted as a function in that it assigns to each index a certain value.

1.2.5   Set operations

You can perform the standard set operations in **ISETL** as in Activity 3. Recall, as you read the following summary, that the convention in this book is that any word in `typeset` font is an **ISETL** key word.

The basic idea of a set is that any object is either in the set or not. You can test for set membership using the operation `in`.

The union (`union`) of two sets is the set of all values which are elements of one set or the other or both.

The intersection (`inter`) of two sets is the set of all elements which are members of both sets.

A set A is a subset (`subset`) of a set B if every element of A is also an element of B.

The difference of two sets (`-`) is the set of all values which are elements of the first set but not the second.

The value in **ISETL** of `{}` is the empty set — the set which has no elements.

The cardinality operator (`#`) applied to a set returns the number of elements of the set.

The operation `pow` applied to a set A constructs the set of all subsets of A, the so-called *power set* of A. Can you see how to figure out the cardinality of the power set?

The operation `arb` applied to a set picks out an arbitrary element of the set and returns it as the result of the operation.

1.2.6   Permutations

Consider a fixed finite set of objects, say the integers $\{1, 2, 3\}$. A particular arrangement of these integers, such as $[3, 1, 2]$ or $[3, 2, 1]$ is called a *permutation* of these integers. Permutations of a fixed set of integers are very important in Abstract Algebra. In this book we will use **ISETL tuples** to represent permutations. Thus, the permutation $[3, 1, 2]$ is represented as `[3,1,2]`. In mathematics, it is common to represent such a permutation as two-row matrix,

$$\begin{pmatrix} 1\ 2\ 3 \\ 3\ 1\ 2 \end{pmatrix}$$

Permutations can also be viewed as functions. For example, if you are working with permutations of the set $\{1, 2, 3\}$ then a particular permutation can be thought of as a function whose domain and range are both this set, $\{1, 2, 3\}$. For example, the permutation, $[3, 1, 2]$ is the function whose value at 1 is 3, at 2 is 1, and at 3 is 2. This is reflected in the matrix representation and also in the **ISETL** access operations. For example, if you do

    p := [3,1,2]; p(2);

You get the answer 1 which is the value of the function p at 2.

In doing more advanced work with permutations, being able to think of them as functions turns out to be very important. We suggest, therefore, that even as you continue to represent them as tuples in your **ISETL** work, you make an effort to train yourself to think of them as functions as well. One way of doing this is by imagining (or writing on paper) the numbers $1, 2, 3$ being added above the given tuple and little arrows pointing down from each number in the top row to the number directly below it.

By the way, how many permutations of the set $\{1, 2, 3\}$ are there?

As a function, a permutation has some very nice properties of which we will be making use throughout this book. For example, such a function is always 1-1, onto, and hence it has an inverse — which is again a permutation. Moreover, since the domain and range of a permutation is the same set, you can always compose two permutations (of the same set). For example, what will **ISETL** return if given the following input?

```
p := [3,1,2];
q := [3,2,1];
[p(q(i)) : i in [1..3]];
```

The first two lines define p and q to be permutations of the set $\{1, 2, 3\}$. The last line is a tuple former. It runs i through the domain $\{1, 2, 3\}$ and for each of these, it evaluates $p(q(i))$. Thus it is computing the composition of the function p with q and the answer is the function pq, represented as a tuple, $[2, 1, 3]$.

1.2.7   Quantification

Quantified logical statements are used in mathematics to express conditions, usually in a definition, statement of a property or a construction. There are two kinds of quantification expressions — *universal quantification* and *existential quantification*. The structure of both have some similarities to the set former.

Following is an example, from Activity 4, of a universal quantification:

```
forall x in Z20 |(x + 0) mod 20 = x;
```

There are two parts to a quantification expression. The first part of a universal quantification begins with the key word forall and is followed by a domain specification as with set formers. The domain specification is completed with the symbol |. The second part is a boolean expression (whose value is true or false).

To evaluate a universal quantification expression, **ISETL** iterates through the values of the variable in the domain specifier. Thus, in this example, it considers every value of x in the set Z20. For each value, the boolean expression is evaluated. If once the value of the boolean expression is false, then the value of the entire universal quantification statement is false. If every value of the variable(s) results in the boolean expression evaluating to true, then the value of the quantification is true.

The existential quantifier is similar, except that it returns true if the value of the boolean expression is true at least once and otherwise, it returns false.

The operation choose is a useful alternative to exists. The syntax is exactly the same, and choose performs the same internal operation as exists. Instead of returning true or false, however, choose will select one value of the variable that makes the condition true and will return that. If there is no such value, then choose returns om. Thus in the case of the following statement from Activity 4

```
e := choose x in Z20 | (forall g in Z20 | (x + g) mod 20 = g);
```

the value of e will be 0.

You may have noticed in Activity 8 something interesting about the behavior of quantification expressions when the negation operation not is applied to them. Consider for example the negation of the **ISETL** statements corresponding to Activities 7(a) and 7(c):

```
not forall x in Z20 | even(x);
not exists x in Z20 | is_tuple(x);
```

It turns out that these two statements are equivalent, respectively, to

```
exists x in Z20 | not even(x);
forall x in Z20 | not is_tuple(x);
```

Moreover, the same kind of transformation always yields equivalent statements. Try to convince yourself that this is indeed the case by either using an intuitive interpretation of these statements, or by trying out some additional examples in **ISETL**, or by manipulating the formal definitions of how quantification and negation expressions are evaluated. Before doing that, however, make sure you write down a careful description of what these transformations are doing.

Using both types of transformations several times over, we can calculate the negation of more complicated expressions. Here, for example, are two equivalent statements corresponding to Activity 7(e):

```
not (forall p in S3 | (exists i in [1,2,3] | odd(p(i))));
exists p in S3 | (forall i in [1,2,3] | not odd(p(i)));
```

(of course, "not odd" could be replaced by "even".)

See if you can play the game with Activity 7(e). You might also derive some fun from recalling (or finding in a Calculus book) the definition of when a function f is continuous at a point a, and formulating what it means for f *not* to be continuous at a.

1.2.8 Miscellaneous ISETL features

There are a number of **ISETL** features that are not extremely important for this course, but we will mention them briefly because they might be encountered occasionally.

We mentioned the operation of union for sets, and its analogue for tuples is *concatenation*. This means tacking one tuple onto the end of another. You can also remove a single element from the beginning or the end of a tuple. Try out the following code.

```
[3,1,2] + [3,2,1];
p := [3,1,2];
take x fromb p; x; p;
take y frome p; y; p;
```

The access operation can be used to change the value of a component of a tuple. The effect of the following line

```
p(2) := 4;
```

is to change the second component of p to the value 4. This only works if p is an expression whose value is a tuple. Sometimes, in order to achieve this, it is necessary to initialize by doing something like p := [ ]; which establiahes p as a tuple by setting it equal to the empty tuple.

Single elements can be placed in or removed from a set. Try out the following code.

```
Z5 := {0..4}; Z5;
Z5 less 0;
Z6 := Z5 with 5; Z6;
```

The **ISETL** value om or OM stands for "no value". This is what is returned if you evaluate a variable to which no value has been assigned.

1.2.9 VISETL

We will occasionally refer in this book to **VISETL** which stands for *virtual ISETL*. By this we mean a programming language which is identical to **ISETL** except that infinite sets are permitted. Otherwise the syntax of the two languages is identical. For example, in **VISETL**, the statements

```
Z := {1,2..};
forall a in Z | a+1 in Z;
```

(which has no meaning in **ISETL**) will assign the (infinite) set of all positive integers to the variable Z and then return the value **true**.

The main difference betwen these two languages is that **ISETL** can be run in your mind and on a computer, but **VISETL** can only be run in your mind.

1.2.10 EXERCISES

1. How many elements are there in the following set? Use the **#** operator in **ISETL** to check your answer.

   ```
   {2, 3, 6 mod 4, {"group", "subgroup", {-1,2..5}},
   "ISETL", {}};
   ```

2. List the elements in each of the following sets and note the number of elements in each. Use **ISETL** to check your answers. If a set is empty, explain why.

   {2..12}
   {4..4}
   {10..1}
   {-2,4..38}
   {0,3..41}
   {0,3..-1}
   {100,90..-5}
   {100,90..100}
   {100,90..101}
   {10,9..0}
   {4,4..8}

3. For each of the following sets of code, predict what result will be returned by **ISETL** and then check your answer on the computer.

   (a)
   ```
   T := {"ring", 3+4, 8};
   7 in T; "r" in T; (1+7) in T;
   "r" + "in" + "g" notin T; 7 notin T;
   T = {7*1, "ring", 15 div 2, "ring"};
   T /= {}; T /= T; {} subset T; T subset T;
   T subset {}; not({"ring"} subset T);
   {7, 1+6, 49 mod 6, 7+0} subset T;
   #(T); pow(T);
   ```

   (b)
   ```
   W := {{"field","group"}, "ring"}; W;
   "field" in W; "ring" in W; {"group"} in W; "group"
   notin W;
   {} in W; {"group", "field"} in W; {} subset W;
   "ring" subset W; {"group", "field", "group"} subset
   W; #(W); pow(W); #(pow(W));
   ```

4. Let Y and Z be defined by

```
Y := { "ideal", 6.9, {2..10}, {{true impl false}, false},
       (10 div -4)+16, {5 in {3,6..9}}, {{}}};
Z := {{false or true}, 28 mod 2, {10,2,9,3,8,4,7,5,6},
       "Ideal", {}, true impl false, {false}, abs(-6.9)};
```

For each expression in the following list, determine if the value is an element of Y, an element of Z, an element of both sets, or neither.

(a) `true`

(b) `false`

(c) `{true}`

(d) `{false}`

(e) `13 + 1`

(f) `{2,3..10}`

(g) `"ideal"`

(h) `{}`

(i) `{{}}`

(j) `{{{}}}`

(k) `{10,9..2}`

5. In the context of the previous exercise, do the following.

   (a) List every expression whose value is in both Y and Z.

   (b) List every expression whose value is in either Y or Z or both.

   (c) List every expression whose value is in Y but not in Z.

   (d) Can you write an **ISETL** expression that will give an answer to any of the above?

6. In the context of the previous two exercises, for each of the following sets, determine whether it is a subset of Y, Z, both Y and Z, or neither Y nor Z.

   (a) `{true}`

   (b) `{false}`

   (c) `{{false}, {true}}`

   (d) `{false, true}`

   (e) `{{2,3..10}, {2..10}, {10,9..2}}`

   (f) `{14, {false and true}, 7-1}`

   (g) `{}`

   (h) `{{}}`

   (i) `{{{}}}`

(j) {10,9..1}

(k) Y

(l) Z

(m) Can you write an **ISETL** expression that will give an answer to any of the above?

7. Try to simplify each of the following expressions to one of S, T, S+T, S*T, {}, or U, where U denotes the set of all **ISETL** values.

   (a) (S*T) + (S*(U-T))

   (b) (S+(T+S)) + (U-T)

   (c) (S+T) + (U-S)

   (d) ((U-S) + T) * (T+S)

   (e) U - (U - (U - ((U-S) + (U-T))))

8. For each of the following **ISETL** set former expressions, give a verbal explanation of the set and then list the elements. For example, if the expression is

$$\{x**2 : x \text{ in } \{2..10\} \mid x \text{ mod } 2 = 0\};$$

then the verbal description might be,

   The set of all squares of the even integers from 2 to 10.

And the list would be

$$\{4, 16, 36, 64, 100\};$$

   (a) {x : x in {2,5..10}};

   (b) {r : r in {2,5..100} | r mod 5 = 0};

   (c) {t**4 + t**2 : t in {-6..6} | even(t div 3)};

   (d) {even(n) : n in {-3,-1..11}};

   (e) {(x*y) mod 3 : x, y in {-8, -7, 0, 7, 8} | x < y};

   (f) {{s,t} : s in {10,8..4}, t in {5..s} | (s+t) mod 2 = 0};

   (g) {(p and q) = (q and p) : p,q in {true, false}};

9. Write **ISETL** set former expressions for each of the following.

   (a) The set of all cubes of those elements of a given set S of integers, which are positive and odd.

(b) The set of all possible sums of two elements, one taken from a given set of integers S and one from a given set of integers T.

(c) The set of all integers between 2 and 100 that are not primes. (Recall that a prime is an integer p which is divisible only by 1,-1, p, $-p$.)

(d) The set of all primes between 2 and 100.

(e) The set of all primes between 2 and 100 that are one more than a multiple of 4.

(f) The set of all integers between 2 and 100 that are the product of two different primes.

(g) The set of all integers between 2 and 100 that are the sum of two different primes.

10. Assume that I is a fixed set of integers and give a verbal description of the set P defined by the following expression.

$$P = \{x : x \in I \mid \text{not}((x < -3) \text{ and } (x \bmod 2 \neq 0))\}$$

11. Assume that S is a **set** and T is a **tuple**, both of which have been previously defined in **ISETL**. Write an **ISETL** expression that will evaluate to a **tuple** whose components are the elements of S and the set whose elements are the components of T.

12. Evaluate the following **tuples** and then use **ISETL** to check your answers.

   (a) [x**2 : x in [1,3..10]];

   (b) [{i,j} : i,j in ["r", "i", "n", "g"] | i /= j];

   (c) [[1..r] : r in [0,2..6]];

   (d) [N+2 < 2**N : N in [0..20]];

   (e) [u*v : u in [-5..0], v in [-5..(u+1)] | (u+v) mod 3 = 0];

13. Assume that T is a **tuple**. Write an **ISETL** expression whose value is the **tuple** of components of T in reverse order.

14. Use the **ISETL** forall, exists, and choose constructs to write code that implements the following statements. Assume that S is the set of multiples of 3 from 0 to 49.

(a) Every odd number in $\{0..50\}$ is in S.

(b) Every even number in S is divisible by 6.

(c) It is not the case that every even number in S is divisible by 6.

(d) There is an odd number in S which is divisible by 5.

(e) There is an even number in S which is divisible by 5.

(f) There is an element m of S such that the number of elements of S that are less than m is twice the number of elements of S that are greater than m.

(g) An element a of S such that for every element x of S, there is an element y of S such that the average of x and y is a.

1.3   Functions in **ISETL**

1.3.1   ACTIVITIES

1. For each of the following sets of **ISETL** code, try to predict what the result would be. Then run the code and check your prediction. Write out a verbal explanation of what the code is doing. Look at the instructions given in Section 1.2.1, Activity 2, p. 12 for an explanation of what is expected in what you write.

   Note: *Each item is considered to be an independent set of code. Therefore, some statements will be repetitions that will not have to be retyped if you ran them earlier and did not leave* **ISETL**.

   (a)
```
f := func(x);
        return (x + 3) mod 6;
     end;
f(5); f(0); f(37);
h := |x -> (x + 3) mod 6|;
h(5); h(0); h(37);
h=f;
forall x in [-10..10] | h(x) = f(x);
```
   (b)
```
fact := func(n);
           return %*[1..n];
        end;
fact(3); fact(5); fact(50);
forall n in [2..20] | fact(n) = n * fact(n-1);
```

```
f := func(x);
        return (x + 3) mod 6;
    end;

f(fact(3));    fact(f(114));
```

(c)

```
!setrandom off
upto := func(n);
            return {1..n};
        end;
upto(1); upto(2); upto(9); upto(10);

n := 100;
upto(n) = upto(n-1) with n;
```

(d)

```
Z12 := {0..11};
H := {0,3,6,9};

coset := func(x);
            return {(x+h) mod 12 : h in H};
        end;
coset(2); coset(5); coset(3); coset(94);
%union{coset(g) : g in Z12};
modH := |x -> {(x+h) mod 12 : h in H}|;
modH(2); modH(5); modH(3); modH(10); modH(94);
```

(e)

```
Z9 := {0..8};
inv := func(x);
            if x in Z9 then
                return
                    choose g in Z9 | (x*g) mod 9 = 1;
            end;
        end;
inv(2); inv(5); inv(6); inv(11); inv(0);
2*inv(2); 5*inv(5); 6*inv(6); 11*inv(11);
```

(f)

```
Z20 := {0..19};
closed := func(H);
            return
                forall x,y in H | (x+y) mod 20 in H;
        end;
closed({0,4..19}); closed({3,5,9}); closed(Z20);
```

2. For each of the following specifications, write an **ISETL** func with the specified input parameters that returns the result of the action that is described. If any auxiliary objects are needed, then construct them as well. Select some specific values and run your func on them to see that it works.

   (a) The input parameter is a single number x and the action is to compute the square mod 20 of the quantity x.

   (b) The input parameter is a single set H of numbers. The action is to check if the result of multiplying a number in H by 3 will always be in H.

   (c) Take H to be the set of those elements of Z12 which are multiples of 4. The input parameter is a single variable x. The action is to check if x is an element of Z12 and, if it is, to return the set of all multiples mod 12 of x by each element of H.

   (d) G is the set Z20 without the element 0. The input parameter is a single variable x and the action is to choose any element of G whose product mod 2 with x is 1.

3. For each of the following sets of **ISETL** code, try to predict what the result would be. Then run the code and check your prediction. Write out a verbal explanation of what the code is doing.

   (a)
```
Z20 := {0..19};
op := func(x,y);
          if (x in Z20 and y in Z20) then
              return (x+y) mod 20;
          end;
      end;
op(3,5); op(9,16); op(4,20);
3 .op 5; 9 .op 16;

Gr := [Z20, op]; Gr(1); Gr(2);
13 .(Gr(2)) 19;
```
   (b)
```
G := {1..12};
o := func(x,y);
          if (x in G and y in G) then
              return (x*y) mod 13;
          end;
      end;
```

```
forall g1,g2 in G | g1 .o g2 in G;
exists e in G | (forall g in G | e .o g = g);
choose e in G | (forall g in G | e .o g = g);
id := choose e in G | (forall g in G | e .o g = g);
id;

forall g in G | (exists g' in G | g' .o g = id);
```

4. Using the following specification, write an **ISETL** func with the specified input parameters that returns the result of the action that is described. If any auxiliary objects are needed, then construct them as well. Select some specific values and run your func on them to see that it works.

> G is the set of integers from 1 to 6 and o is the operation that takes two variables and computes their product mod 7. The input parameter to your func is a single variable g. Your func checks that g is an element of G and, if it is, chooses an element of G whose "product" with g under the operation o is equal to 1.

You may wish to keep this code for future use.

5. Write the same code as in Activity 3(b) with 12 replaced by 5 and 13 replaced by 6. You may find it convenient here to use the copy feature on your computer.

6. In Activity 3(b) and Activity 5 you did exactly the same thing twice with slight changes in the numbers.

   What remained the same and what changed? Write out an explanation of all similarities and differences.

7. Write **ISETL** funcs with the given names according to each of the following specifications. In each case, set up specific values for the parameters and run your code to check that it works.

   (a) The func is_closed has two input parameters: a set G and a func o which is some operation on two variables from G such as the one in Activity 3. The action of is_closed is to determine whether the result of o, when applied to two elements of G, is always an element of G. This is indicated by returning the value true or false.

   (b) The func is_commutative has two input parameters: a set G and a func o which is some operation on two variables from G such as the one in Activity 3. The action of is_commutative is to determine whether or not the result of o depends on the order

of the elements. This is indicated by returning the value true
or false.

(c) The func identity has two input parameters: a set G and a
func o which is some operation on two variables from G such as
the one in Activity 3. The action of identity is to search for
an element e of G which has the property that for any element
g of G, the result of the operation o applied to e and g is again
g. This is indicated by returning the value of e if it exists or om
if it does not.

(d) The func inverses has three input parameters: a set G, a binary operation o on G, and an element g of G. We assume that
G and o are such that identity(G,o) is defined (doesn't return om). The action of inverses is to first assign the value of
identity(G,o) to a variable e, then search for an element g' of
G which has the property that the result of the operation o applied to g' and g is e. The func inverses returns the element
g' if it exists or om if it doesn't.

8. For each of the following sets of **ISETL** code, try to predict what the
result would be. Then run the code and check your prediction. Write
out a verbal explanation of what the code is doing.

(a)
```
p := {[1,3], [2,4], [3,2], [4,1]};
p(1); p(3);
pt := [p(i) : i in [1..4]]; pt;
pt(1); pt(3);

q := {[3,1], [4,2], [2,3], [1,4]};
pq := {[i,p(q(i))] : i in [1..4]};
pqt := [p(q(i)) : i in [1..4]];
```
(b)
```
next := func(n);
            if n in [1..10] then return n+1; end;
        end;

mnext := {[n,n+1] : n in [1..10]};

next(6); mnext(6); next(9); mnext(9); next(11);
mnext(11);

forall n in [0..11] | mnext(n) = next(n);
domain(mnext); domain(next);
```

(c)
```
G := {1..12};
o := func(x,y);
        if (x in G and y in G) then
            return (x*y) mod 13;
        end;
    end;


m13 := {[[x,y], x .o y] : x,y in G};
m13;
m13(3,5); m13(2,4);
3 .m13 5; 2 .m13 4;
forall x,y in G | x .m13 y = x .o y;
```

9. Describe what the following code is doing.
```
SetNot := proc(pair);
            G := pair(1); o := pair(2);
            e := choose x in G | (forall g in G | x .o g
            = g);
            inv := {[g, choose g' in G | g' .o g = e] :
            g in G};
        end;


pair := [ ];
pair(1) := {0..5};
pair(2) := func(x,y);
            if (x in G and y in G) then
                return (x*y) mod 6;
            end;
        end;


pair;
SetNot(pair);
G; e; inv(5); inv(2);
```

What would happen if, after running this code (and defining the funcs in Activity 7), you entered statements such as 3 .o 5, is_closed(G, o), is_commutative(G,o), identity(G,o)?

Can you imagine why we might want to go to all the trouble of something like SetNot?

10. An **ISETL** func can also return a func. Look at the following code and write down an explanation of what it does:

```
add_a := func(a);
        return
            func(x);
                return x + a;
            end;
        end;
```

Predict the results of the following **ISETL** statements:

```
add_a(3); add_a(3)(5); add_a(3)(2); add_a(7)(4);
f3 := add_a(3); f3(5); f3(2);
f7 := add_a(7); f7(4);
```

11. Write down an **ISETL** func compose which takes as input two **ISETL** funcs f and g and returns a func representing their composition; that is, the func compose returns a func which for each input x returns f(g(x)). Use your func to compose some of the funcs previously defined in the Activities, and study the resulting new funcs.

1.3.2   Funcs

Look, for example, at the first three lines of code in Activity 1. This is an assignment statement which tells **ISETL** to figure out what is meant by the code

```
func(x);
    return (x+3) mod 6;
end;
```

and assign it to the variable f. Thus, **ISETL** will know in the future that the *value* of f is the *meaning* of this code.

Now look at the first expression in the fourth line of code, f(5);. This tells **ISETL** to perform the operations that it worked out for the func that was assigned to f. It also tells **ISETL** that, in doing these operations, the value of x is to be taken as 5. Thus, it computes the value (5+3) mod 6, which is 2, and the return statement tells it to return that value.

There are several ways to represent functions in **ISETL**. You have seen that a tuple can be interpreted as a function and this idea is used in working with permutations (see Section 1.2.6, p. 17.) The most important structure for representing functions in **ISETL** is the func. A func has three parts: a header line which consists of the key word func followed by a list of input parameters enclosed in parentheses and separated by commas; a body, which is a sequence of **ISETL** statements; and a closing line consisting of the key word end. The body of a func may contain any number (including 0) of *return* statements. A return statement consists of the key word **return** followed by an expression which evaluates to some **ISETL** value.

Because a func is generally constructed for repeated use, it is usually given a name by assigning it to be the value of some variable. When **ISETL** encounters a set of code which it recognizes as a func, it performs whatever internal operations are necessary for it to be able (at some other time) to perform the statements in the body of the func any time the func is called.

Whenever the action of a func comes to a **return** statement, the expression is evaluated, the operation of the func is terminated, and the value of the expression is returned. Although a func can have more than one return statement, each call to a func will only make use of one of them because the operation is terminated once a return statement has been completed. If a func has no return statement, or gets to its end without encountering one, then it will return the value om.

A func can have any number of input parameters (see, for example Activity 3) and these parameters can take on any **ISETL** values.

A func can only return a single value (the result of evaluating the expression in a return statement or, if none is encountered, om). If you would like to get more than one value from a func, put them in a **tuple** to make a single value and return the **tuple**.

1.3.3 Alternative syntax for funcs

There is an alternative syntax that can be used for funcs which do nothing but compute the value of an expression (like f in Activity 1(a), but not inv in 1(e).) In this version, the entire func definition is contained within the two delimiters | and |. The definition begins with a variable. It is followed by the symbol -> which is followed by an expression. This syntax is essentially a shorthand. For example, the action of the func h in Activity 1(a) is identical to that of the func f in that activity. Thus, the functions they represent are equal even though, as you saw, the expression h=f returns false.

One reason why this syntax is used is because it is very close to a shorthand that mathematicians may use. For example, in defining a function such as the one implemented by the func h, it is possible to write

$$h : x \mapsto (x + 3) \bmod 6$$

(read *h is the function defined by x goes to* $(x + 3)$ mod 6.) Compare this with the **ISETL** syntax,

```
h := |x -> (x + 3) mod 6|;
```

1.3.4   Using funcs to represent situations

One of the things that funcs are used for is to represent situations. Look at Activity 2(c), for example. The first sentence says that Z12 and H must be defined. These are statements that come before the definition of the func begins. Now you are ready to define the func. You have to decide if you are going to assign it to a variable and what name to use. The second sentence of the activity tells you that there is to be a single input parameter and it is suggested that its name be x.

The last sentence of the activity describes the action. The action is to be embedded in an if...then... statement. The test is for x to be an element of Z12. If so, the action is to construct the set of all multiples x*h of x by each element h of H. This set is returned. If the value of x was not an element of Z12, the activity does not say what is to be returned so it is understood that the func returns om in this case.

1.3.5   Funcs for binary operations

When a func has two input parameters, which are assumed to represent two elements from some set, and the returned value is also assumed to belong to the same set, then the func may be said to represent a *binary operation*. Binary operations are extremely important in Abstract Algebra and you will spend a great deal of time working with them.

In mathematics, it is the usual practice to write the name of a binary operation *between* the two parameters, rather than before them. Thus, we write $a+b$ rather than $+(a, b)$. We can do something very similar in **ISETL**. If o is any binary operation then, instead of o(e,g) we have the option of writing e .o g. Putting the period before the name of the operation is the signal to **ISETL** that the operation is between the two parameters, not before them.

1.3.6   Funcs to test properties

The next chapter, which introduces the study of groups, begins with a consideration of several funcs that test binary operations for various properties. We have done some of them here. Look, for example, at Activity 7(a). The func you are to write has two input parameters, a set G and a binary operation o. It is assumed, of course, that before you call this func you will have defined a set and a binary operation. The definition of the func will contain a single boolean expression very similar to one of the lines of code in Activity 3(b). The value of this expression is returned as the result of a call to this func.

1.3.7 Smaps

In activity 8 you compared two **ISETL** objects: the `func` `next`, which assigns to each number in [1..10] the next larger number, and the set of ordered pairs

$$\texttt{mnext := \{ [n,n+1] : n in [1..10]\};}$$

Hopefully you also discovered that expressions like `mnext(6)` = `next(6)` return `true`.

In **ISETL**, an ordered pair is a tuple with only the first two components defined. A set of ordered pairs is called a `map`. The `map` gives us one more way to represent functions in **ISETL** — as in mathematics. As you can see in the above example, the way a `map` represents a function is that it assigns the second component of each ordered pair to the first component of the same pair. However, not every `map` represents a function. Since a function assigns to each element in its domain a *single* value, a `map` representing a function must have the following additional property:

> *No value appears as a first component of more than one pair in the map.*

A `map` with this property is called an `smap` (single-valued map). Any time an `smap` is constructed and assigned to a variable, that variable can be used as a function.

For the purpose of representing functions, the `func` in **ISETL** has the advantage of explicitly showing the process by which the returned value is obtained from the input value. An `smap`, on the other hand, represents more accurately the mathematical notion of when two functions are to be considered equal. Recall for example the `funcs` f and h of Activity 1(a) which were virtually identical, but were not considered as such by **ISETL**. An `smap` is also faster to evaluate on the computer than a `func`. In the next chapters we will occasionally use this fact in some calculations that take a long time to perform.

Let us look at another example, the one appearing in Activity 8(c). There the variable m13 is set to have the value of an `smap`. The pairs are a little complicated. The first component of each pair is itself a pair. The second component is the result of multiplying, mod 13, the two numbers in the first component.

Thus, the expression m13(3,5) is evaluated as follows. The two values 3, 5 are put together to form a pair [3,5]. This is the value on which the function represented by the `smap` must operate. The operation consists of looking up the pair [3,5] in the set of ordered pairs to find one whose first component has the value [3,5]. Because of the requirement for `smaps` there will be only one. The second component of this pair is 2 and this is the value of m13(3,5).

1.3.8 *Procs*

Activity 9 has an example of an **ISETL** proc or *procedure*. A procedure is the same as a func except that it has no return statement and does not return a value. It is used to perform some internal operations such as establishing the values of certain variables. It is also used for external effects on the screen (such as printing or drawing something), or other devices (disk, printer, and so forth).

1.3.9 *EXERCISES*

1. Each of the following is a description of a function. Use an **ISETL** func to implement it, with a restricted domain where appropriate. Calculate the value of the function on at least three values of the domain in two ways: first with paper and pencil, using the given verbal description and second, on the computer, using your func. Explain any discrepancies.

   (a) The function takes a tuple of length 4 whose components are positive integers and computes the sum of the cubes of the components.

   (b) The domain of the function is the set of rational numbers (you will have to do !rational on in this func). The function transforms a rational number x into $x-1$ if $0 < x \le 2$ and into $x^2 + 1$ otherwise.

   (c) The domain of the function is the set of rational numbers (you will have to do !rational on in this func). The function transforms a rational number x into $x-1$ if $0 \le x < 2$, while the other values are determined by the condition that if x is increased by 2, then the value of the function is doubled.

   (d) The domain of the function is the set of rational numbers (you will have to do !rational on in this func). The name of the function will be g and its values are determined by

   $$g(0) = 3, \quad g(x) = |x| \text{ for } 1 \le |x| \le 3, \quad g(3x) = g(x)$$

   where the last relation is to hold for all x.

   (e) The domain of the function is the set of all points of the square centered at the origin of a rectangular coordinate system with sides parallel to the coordinate axes and having length 2. The action of the function is to rotate the square counterclockwise through an angle of 180°.

   (f) The domain of the function is the set of all points of the square centered at the origin of a rectangular coordinate system with sides parallel to the coordinate axes and having length 2. The action of the function is to flip the square over along its diagonal from the upper left to the lower right vertex.

(g) The function determines the number of non-om components of a `tuple`. (Note that not all of the components of a `tuple` between the first and last must be defined).

2. Write `funcs` `a20` and `m20` each of which has domain $Z20 = \{0..19\}$ and which implement addition and multiplication mod 20.

3. Write **ISETL** `funcs` with the given names according to each of the following specifications. In each case, set up specific values for the parameters and run your code to check that it works.

   (a) The `func` `is_associative` has two input parameters: a set `G` and a `func` `o` which is some operation on two variables with values in `G` such as the ones in the previous exercise. The action of `is_associative` is to determine whether the operation represented by `o` is associative. This is indicated by returning the value `true` or `false`.

   (b) The `func` `is_identity` has three input parameters: a set `G` and a `func` `o` which is some operation on two variables with values in `G` such as the ones in the previous exercise, and a single element of `G`. The action of `is_identity` is to determine whether the element has the property that if the operation is applied to it and any other other element `x` of `G`, then the result is again `x`. This is indicated by returning the value `true` or `false`.

   (c) The `func` `is_inverse` has five input parameters: a set `G` and a `func` `o` which is some operation on two variables with values in `G` such as the ones in the previous exercise, and three single **ISETL** values. The action of `is_inverse` is first to determine whether `is_identity` applied to the first of the three values returns `true` and, if so, if the operation applied to the other two values gives a result equal to the first value. This is indicated by returning the value `true` or `false`.

4. What will appear on the screen after the following lines are entered in **ISETL**? Predict the answer, run it and explain why your answer agreed or disagreed with what the computer gave.

```
F := func(M,N);
        return {1,N..M} + {-M,-N..1};
     end;

F(2,4); F(2,3); F(3,2);
```

5. Suppose the following has been run in **ISETL**.

```
F := func(P,Q);
        return P impl Q;
     end;

G := func(P,Q);
        return (P and Q) or P;
     end;
```

Answer the following questions, and give full explanations of the results.

(a) What is the meaning of the following two assignments?

   i. S := {F};
   ii. T := {F,G};

(b) What will appear on the screen if, after the two assignments are made, the following is run?

```
for x in T do
       print x(true, true);
       print x(false, true);
    end;
```

(c) After the two assignments are made, what will appear on the screen if the following two pieces of code are run?

   i. arb(S)(true, false);
   ii. arb(T)(false, true);

(d) Write out an explanation of the **ISETL** operation arb.

6. Write an **ISETL** map that implements the function whose domain is Z_{20} and asssigns to each element x an element y such that $x + y$ mod $20 = 0$. Is this an smap? Explain.

7. Do the same as the previous exercise with addition replaced by multiplication and 0 replaced by 1.

8. Write an **ISETL** smap that implements the operation of addition mod 20 in Z_{20}.

9. Write an **ISETL** func that accepts a pair consisting of a set G and an operation o that takes two elements of G and returns an element of G. The action of the func is to convert this pair into an smap which implements the operation. Use your func to do the previous exercise.

2
Groups

2.1 Getting acquainted with groups

2.1.1 ACTIVITIES

1. Following is a list of some funcs that you worked with in the first
   chapter.

   ```
   is_closed
   is_associative
   is_identity
   is_inverse
   is_commutative
   has_identity
   has_inverses
   ```

   Notice that the funcs identity and inverse are changed to has_
   identity and has_inverses to indicate that they are to return only
   the boolean value true or false.

   Write a description of what each func does including the kind of
   objects accepted, what is done to them and the kind of object that
   is returned.

2. For each of the following, construct an **ISETL** pair of objects G and op where

   (a) G is Z12 (the integers mod 12) and op is a12 (addition mod 12).

   (b) G is Z12 (the integers mod 12) and op is m12 (multiplication mod 12).

   (c) G is twoZ12 (the even integers mod 12) and op is m12.

   (d) G is Z12 - {0} and op is m12.

   (e) G is Z5 (the integers mod 5) and op is m5 (multiplication mod 5).

   (f) G is Z5 - {0} and op is m5.

   (g) G is S3 (the set of permutations of {1, 2, 3}) and op is composition of permutations.

   For example, in (a) you could write code such as the following.

   ```
   Z12 := {0..11};
   a12 := | x,y -> (x + y) mod 12|;
   ```

3. Guess what the func has_identity will return when applied to each of the pairs of Activity 2. Check your guess on the computer. Guess what the func has_inverses will return when applied to each of the pairs of Activity 2.

4. Apply each of your funcs from Activity 1 to each pair you set up in Activity 2. List the properties which are satisfied by all seven of the pairs. Here and in the following activities, ignore those funcs which refer to single elements rather than the whole set.

5. In this Activity, you are to investigate how the result of Activity 4 changes if some of the examples are removed from the list in Activity 2. For each of the following items, perform the indicated removals and then see which of the properties are satisfied by all of the remaining pairs.

   (a) Remove Z12 - {0} from the list in Activity 2.

   (b) Remove Z12 - {0} and twoZ12 from the list in Activity 2.

   (c) Remove Z5, Z12 - {0}, twoZ12, and the second occurence of Z12 (the one that goes with m12) from the list in Activity 2.

   (d) Remove Z5, Z12 - {0}, twoZ12, the second occurence of Z12 and S3 from the list in Activity 2.

6. Consider a square made of thin metal lying on a table. Label the corners. Imagine picking up the square and putting it back in exactly the same spot. The resulting situation looks just as it did in the beginning, but the vertices could possibly be in different positions. Call this transformation of the square a repositioning or symmetry. One can imagine doing this twice in succession, perhaps using a different transformation the second time. The result of performing two successive repositionings will also be a repositioning. This may be considered as an operation which takes two symmetries and produces a symmetry by doing first one and then the other.

   (a) How many different symmetries does a square have?

   (b) Explain how a symmetry may be considered to be a permutation.

   (c) Consider an **ISETL** representation of the set of symmetries of a square and also of the operation. Make an **ISETL** pair as you did in Activity 2 (See the explanation there.).

   (d) Which of the properties in Activity 1 are satisfied by the symmetries of the square?

   Note: Some of these tasks can take a very long time, depending on the speed of the machine.

7. Here is a list of some more sets with binary operations. Construct an **ISETL** pair for each of them as you did in Activity 2. Which of the properties in Activity 1 are satisfied by these pairs?

   Remark. Some of the following systems can only be constructed and run in **VISETL**. This means that all of your work must be done by hand and in your mind.

   (a) G is S4 (the set of permutations of $\{1, 2, 3, 4\}$) and op is composition of permutations.

   (b) G is the set $\{1, -1\}$ of two integers and op is ordinary multiplication.

   (c) G is Z (the set of all integers — positive, negative and zero) and op is addition.

   (d) G is Z $-\{0\}$ and op is multiplication.

   (e) G is 2Z (the set of all even integers) and op is addition.

   (f) G is 2Z (the set of all even integers) and op is multiplication.

   (g) G is Q (the set of all rational numbers) and op is addition.

   (h) G is Q $-\{0\}$ and op is multiplication.

8. Write a func is_group which accepts a set and a binary operation on that set and tests whether the properties in the lists that you obtained as your answers to Activity 5 are satisfied.

9. Let S be a set with a binary operation and suppose that the result of the operation applied to $a, b \in S$ is written ab. What would you imagine is meant by a^2, a^3, a^{-2}, a^0? You may need to assume that some of the properties listed in Activity 1 are satisfied.

10. In the context of Activity 9, look through the various binary operations you have constructed and investigate the relation $(ab)^{-1} = a^{-1}b^{-1}$. Is it always true? Sometimes? Never? Can you prove anything? Can you think of any modification that would make it always true?

11. The same as the previous activity for the relation $(ab)^2 = a^2b^2$.

2.1.2 Definition of a group

In the activities at the beginning of this section you constructed many mathematical systems and examined their properties. More specifically, you constructed certain sets, defined operations on them and studied various properties of these sets under the operations.

For example, you constructed the set S_3 of all permutations of $\{1, 2, 3\}$, defined the "product" of any two such permutations and checked that this operation satisfies the properties of closure, associativity, existence of an identity element, and the existence of an inverse permutation for every permutation. It does not, however, satisfy the property of commutativity.

The four properties which do hold for S_3 are also satisfied by many other important mathematical systems. Rather than study each system separately, we are going to consider these four properties, and investigate what facts can be determined just from them alone. This approach, called the axiomatic or abstract approach, is very powerful and adds much to the clarity and generality of investigations into mathematical systems.

Each of the systems in the activities consists of two things: a set S and a binary operation $\circ$. By a binary operation on a set S, we mean an operation that assigns, to each ordered pair $[a, b]$ of elements in S, a unique element of S (denoted $a \circ b$).

Definition 2.1 (Group.) *A set G together with a binary operation $\circ$ is called a group (relative to the operation $\circ$) if the following four properties (called the group axioms) are satisfied:*

Closure. *For all $a, b \in G$, $a \circ b \in G$.*

Associativity. *For all $a, b, c \in G$, $a \circ (b \circ c) = (a \circ b) \circ c$.*

Existence of identity. *There exists an element $e \in G$ (called an identity element) such that for all $a \in G$, $e \circ a = a \circ e = a$.*

Existence of inverses. *For all $a \in G$, there exists an $a' \in G$ (called an inverse of a) such that $a \circ a' = a' \circ a = e$.*

If, in addition, G satisfies[1]
Commutativity. *For all $a, b \in G$, $a \circ b = b \circ a$*

then it is called a commutative *or* Abelian *group (after the Norwegian mathematician Nils Abel).*

Thus, S_3 and Z_{12} are groups and moreover, Z_{12} is a commutative group.

2.1.3 Examples of groups

We are now going to examine some of the examples you worked with in the activities to determine whether or not they are groups. In cases where it is a group we will prove that fact. Where it is not, we will see just which group axioms are violated. The results are summarized at the end of the chapter (p. 80).

The examples we consider fall naturally into four categories: number systems, integers mod n, permutations (symmetric groups), and geometric transformations (symmetries).

Number systems

Proposition 2.1 *The integers with addition $[Z, +]$ form an Abelian (commutative) group.*

To show this, we need only go carefully through the list of axioms for a group, each time citing an appropriate well-known property of the integers.

Closure. The sum of any two integers is an integer;

Associativity. For all $a, b, c \in Z$, $(a + b) + c = a + (b + c)$;

Existence of identity. The number 0 is an identity element since it satisfies $a + 0 = 0 + a = a$ for all $a \in Z$;

Existence of inverses Given an integer $a \in Z$, $-a$ is also in Z. Moreover, $-a$ is the inverse of a since $a + (-a) = (-a) + a = 0$;

Commutativity. For all $a, b \in Z$, $a + b = b + a$.

---

[1]Strictly speaking, we should refer here to the set G together with the operation $\circ$, rather than the set G alone. For simplicity however, we will often use this informality.

Note: Since commutativity and associativity of addition and multiplication are properties of all numbers, we will take these properties for granted when dealing with number systems.

Proposition 2.2 *The integers with multiplication* $[\mathcal{Z}, *]$ *do not form a group.*

Proof. The (only) axiom that fails here is the existence of inverses for all elements. To see this, we first note that 1 is the multiplicative identity $(a * 1 = 1 * a = a)$[2]. Since $a * 0$ is always equal to 0 rather than 1, we conclude that 0 cannot have an inverse. Hence, $[\mathcal{Z}, *]$ is not a group.

$\square$

Remark. (non-invertibility of 0.) What we have seen is a general phenomenon: Any number system containing 0 (except for the trivial case $\{0\}$, in which 0 is the only element of the set) cannot be a multiplicative group since 0 cannot have an inverse.

Returning to our example $[\mathcal{Z}, *]$, we might try to fix matters by removing zero. However, the number 2 (or any other integer except 1 or -1) also does not have an inverse, since there is no integer a such that $2 * a = 1$. Hence even the remaining system $[\mathcal{Z}-\{0\}, *]$ is not a group. The system $[\mathcal{Z}-\{0\}, *]$ does satisfy all the other group axioms. In particular, it is closed since the product of two non-zero integers is again a non-zero integer.

Note that we actually know a number a such that $2 * a = 1$, namely the fraction, or rational number, $\frac{1}{2}$. In this sense, $\frac{1}{2}$ is a multiplicative inverse of 2. However, $\frac{1}{2}$ is not in $\mathcal{Z}$. The next example is a follow-up to this observation.

Proposition 2.3 *The non-zero rationals with multiplication* $[\mathcal{Q} - \{0\}, *]$ *form an Abelian group.*

Proof. Recall that a number is called rational if it can be expressed as a quotient a/b, where $a, b \in \mathcal{Z}$ and $b \neq 0$. The elements of $\mathcal{Q} - \{0\}$ are, therefore, precisely those numbers that can be expressed as a/b with $a, b \in \mathcal{Z} - \{0\}$. In particular, all integers are rational numbers since an integer a can always be written as $a/1$.

Closure. We want to show that the product of two non-zero rational numbers is again a non-zero rational number. This follows from the formula for multiplying rational numbers:

$$\frac{a}{b} * \frac{c}{d} = \frac{a * c}{b * d}$$

Here a, b, c, d are in $\mathcal{Z} - \{0\}$, hence so are $a * c, b * d$.

---

[2]Strictly speaking, this proof requires the fact that 1 is the *only* identity element of $[\mathcal{Z}, *]$.

Associativity and commutativity. These again are standard properties of numbers.

Identity. 1 is a rational number and is a multiplicative identity.

Inverse. Given $a/b \in \mathcal{Q} - \{0\}$, the rational number b/a is also in $\mathcal{Q} - \{0\}$, and is an inverse of a/b because

$$\frac{a}{b} * \frac{b}{a} = \frac{a * b}{b * a} = 1$$

$\Box$

Integers mod n

In the activities preceding this section you worked with various systems of modular arithmetic such as $[\mathcal{Z}_{12}, +_{12}]$, $[\mathcal{Z}_{12}, *_{12}]$, $[\mathcal{Z}_{12} - \{0\}, *_{12}]$, and $[\mathcal{Z}_5 - \{0\}, *_5]^3$. Unlike the previous category of number systems, these are all finite and can be run in **ISETL**. In fact, one of your **ISETL** activities was to check which of these systems forms a group. (Applying is_group to these four should have resulted in true, false, false, and true, respectively). We will now look at these four examples and see just why they are or are not groups. In the next section, we will introduce the general class of such examples more systematically and develop theoretical tools for determining whether or not any system of integers mod n with addition or multiplication mod n forms a group.

Proposition 2.4 *The integers mod 12 with addition mod 12, $[\mathcal{Z}_{12}, +_{12}]$, form an Abelian group.*

Proof. We will run through the axioms for a group, citing appropriate properties of addition mod 12. Recall that $\mathcal{Z}_{12} = \{0, ..., 11\}$ and that the operation $+_{12}$ (addition mod 12) on two elements of $\mathcal{Z}_{12}$ consists of first adding them and then dividing the sum by 12 and taking only the remainder. Thus $7 +_{12} 8 = (7 + 8) \bmod 12 = 15 \bmod 12 = 3$.

Closure. Since the remainder is always between 0 and 11, the result of the operation is in $\mathcal{Z}_{12}$.

Associativity. One way of doing this is to check all possibilities, which is what is_group does. A more generally applicable method begins with the observation that, just as $a +_{12} b$ means $(a + b) \bmod 12$, it turns out that $(a +_{12} b) +_{12} c$ means $((a + b) + c) \bmod 12$. Similarly, $(a +_{12} (b +_{12} c))$ means $(a + (b + c)) \bmod 12$. This is discussed more

---

[3]Note the slight change in notation. In mathematical text we can use a much larger set of symbols than in **ISETL**. Thus we have used here the more suggestive $+_{12}$ instead of **ISETL**'s a12 to denote addition mod 12, etc.

fully in the next section, but for now let's look at an example. Suppose $a = 7, b = 8$, and $c = 10$. Then

$$
\begin{aligned}
(a +_{12} b) +_{12} c &= (7 +_{12} 8) +_{12} 10 \\
&= (15 \bmod 12) +_{12} 10 \\
&= 3 +_{12} 10 \\
&= 13 \bmod 12 \\
&= 1
\end{aligned}
$$

$$
\begin{aligned}
((a + b) + c) \bmod 12 &= ((7 + 8) + 10) \bmod 12 \\
&= 25 \bmod 12 \\
&= 1.
\end{aligned}
$$

We can now show that associativity of addition mod 12 is "inherited" from the associativity of regular addition of integers. Indeed we have

$$
\begin{array}{lll}
(a + b) + c & = a + (b + c) & \text{(associativity of} \\
((a + b) + c) \bmod 12 & = (a + (b + c)) \bmod 12 & \text{ordinary} \\
& & \text{addition)} \\
(a +_{12} b) +_{12} c & = a +_{12} (b +_{12} c) & \text{(above} \\
& & \text{observation)}
\end{array}
$$

Existence of identity. The number 0 is an identity.

Existence of inverses. The inverse of a is $12 - a$ (Why?).

Commutativity. Inherited from ordinary addition (Exercise 1).

$$\square$$

Proposition 2.5 *The integers mod 12 with multiplication mod 12, $[\mathcal{Z}_{12}, *_{12}]$, do not form a group.*

Proof. The only possible identity is 1. But then 0 does not have an inverse because multiplying any number by 0 gives 0, which is different from 1 (even in mod 12). This is another example of the non-invertibility of 0 which we mentioned in relation to number systems.

$$\square$$

Proposition 2.6 *The integers mod 12 without 0 with multiplication mod 12, $[\mathcal{Z}_{12} - \{0\}, *_{12}]$, do not form a group.*

Proof. Since $2 *_{12} 6 = 0$, this system is not closed. It is also possible to deduce from this relation that 2 and 6 are not invertible. (Exercise 2).

$$\square$$

Proposition 2.7 *The integers mod 5 without 0 with multiplication mod 5,* $[\mathcal{Z}_5 - \{0\}, *_5]$ *, form an Abelian group.*

Proof.

Closure. The remainder on division by 5 of the product of two of the integers 1,2,3,4 is always between 0 and 4. The product of any two of the numbers 1,2,3,4 is never divisible by 5 so the remainder on dividing by 5 can never be 0; thus it must be one of 1,2,3,4.

Associativity and commutativity. These are handled in a manner similar to that for $\mathcal{Z}_{12}$.

Existence of identity. 1 is an identity.

Existence of inverses. The inverses of 1,2,3,4 are 1,3,2,4 respectively. (Again, a general method for proving the existence of inverses in this situation will be shown in the next section.)

▯

Symmetric groups

In the activities you worked with the set S_3 of permutations of $\{1, 2, 3\}$ together with the operation *os*, the composition of permutations. You also worked to a lesser extent with the set S_4 of all permutations of $\{1, 2, 3, 4\}$ together with the operation *os*. In both cases, the **ISETL** func is_group reports that the system is a group. It is possible, however, that you found is_group to take too long for S_4 and so maybe you never got an answer for that case. In this section we will look at the system $[S_4, os]$ and see that it is a group. Since we will only consider one operation on S_4 we will sometimes refer to the system as just S_4. In the next section the general case of S_n will be treated.

Representations of S_4. Intuitively, a permutation of $\{1, 2, 3, 4\}$ can be thought of as reordering these numbers. In **ISETL** we represented each permutation as a tuple. Thus, [2,4,1,3], [3,1,2,4] and [1,2,3,4] are permutations in S_4. Note that by "reordering" we mean "the same numbers, possibly in a different order". In other words, all the items of the tuple should be different from each other and all four should be present. Thus, [1,1,2,4] is not a permutation both because 1 appears twice and because 3 is absent. Note also that because we are dealing with a finite set, each of the two requirements ("all different" and "all appear") actually implies the other (exercise). Do you see how a permutation of $\{1, 2, 3, 4\}$ can be defined as any tuple [a,b,c,d] where a,b,c,d in $\{1,2,3,4\}$ and #({a,b,c,d}) = 4? Compare this with the definition of S3, p. 12.

Permutations as functions. When we think of a permutation, or re-ordering, we think of things (numbers, in this case) and their places. For more advanced work with permutations, it is convenient to display this information more explicitly. Thus the standard notation for the permutation that we represented by the tuple p = [3,1,2,4] will be

$$p = \left(\begin{array}{cccc} 1\,2\,3\,4 \\ 3\,1\,2\,4 \end{array} \right).$$

One advantage of this notation is that, while it is easy to continue thinking of permutations as tuples (by simply ignoring the top row), it now becomes easier to think of them as functions , i.e. p is the function that maps the numbers in the top row to the number directly below them in the bottom row.

More explicitly, this is the function $p : \{1, 2, 3, 4\} \rightarrow \{1, 2, 3, 4\}$, defined by $p(1) = 3$, $p(2) = 1$, $p(3) = 2$, $p(4) = 4$.

Note that the notation $p(i)$ is meaningful whether we think of the permutation p (or represent it in **ISETL**) as a tuple or as a func, and in fact will have the same value in both cases.

Permutation product as composition. One of the main benefits of viewing permutations as functions is that the product of permutations becomes nothing more than composition of functions. If p and q are permutations in S_4, their product, written pq, is given by

$$pq(i) = p(q(i)), \quad i = 1, 2, 3, 4.$$

In other words, to obtain the product pq, first do q, then (on the result) do p. For example,

$$\left(\begin{array}{cccc} 1\,2\,3\,4 \\ 3\,2\,4\,1 \end{array} \right) \left(\begin{array}{cccc} 1\,2\,3\,4 \\ 4\,1\,2\,3 \end{array} \right) = \left(\begin{array}{cccc} 1\,2\,3\,4 \\ 1\,3\,2\,4 \end{array} \right).$$

This is because in the product we have $1 \rightarrow 4 \rightarrow 1$, $2 \rightarrow 1 \rightarrow 3$, $3 \rightarrow 2 \rightarrow 2$ and $4 \rightarrow 3 \rightarrow 4$. (Remember that the right-hand factor is done first.)

Proposition 2.8 *The permutations of $\{1, 2, 3, 4\}$ with permutation product, os, form a group.*

Proof.

Closure. If you reorder the numbers 1,2,3,4 in some way and then reorder the result, then altogether what you have done still amounts to some reordering of 1,2,3,4.

Associativity. Composition of functions is associative. If p, q, r are permutations, then for any $i = 1, 2, 3, 4$ we have

$$((pq)r)(i) = pq(r(i)) = p(q(r(i))) = p(qr(i)) = (p(qr))(i).$$

Existence of identity. The permutation $\begin{pmatrix} 1\,2\,3\,4 \\ 1\,2\,3\,4 \end{pmatrix}$ is an identity.

In fact, this permutation leaves everything exactly as it is, so if you apply it to the result of some permutation, it is no different than just applying the original permutation.

Existence of inverses. If you reorder 1,2,3,4 you can always undo the reordering: just send each number to where it originally came from. This "undoing" is the inverse of the permutation. This is because doing and then undoing is the same as doing nothing, which is the identity permutation. In terms of the standard notation, the inverse of $\begin{pmatrix} 1\,2\,3\,4 \\ a\,b\,c\,d \end{pmatrix}$ is $\begin{pmatrix} a\,b\,c\,d \\ 1\,2\,3\,4 \end{pmatrix}$ since their product gives the identity.

For example, the inverse of $\begin{pmatrix} 1\,2\,3\,4 \\ 3\,2\,4\,1 \end{pmatrix}$ is $\begin{pmatrix} 3\,2\,4\,1 \\ 1\,2\,3\,4 \end{pmatrix}$ or $\begin{pmatrix} 1\,2\,3\,4 \\ 4\,2\,1\,3 \end{pmatrix}$. Indeed,

$$\begin{pmatrix} 1\,2\,3\,4 \\ 4\,2\,1\,3 \end{pmatrix}\begin{pmatrix} 1\,2\,3\,4 \\ 3\,2\,4\,1 \end{pmatrix} = \begin{pmatrix} 1\,2\,3\,4 \\ 3\,2\,4\,1 \end{pmatrix}\begin{pmatrix} 1\,2\,3\,4 \\ 4\,2\,1\,3 \end{pmatrix} = \begin{pmatrix} 1\,2\,3\,4 \\ 1\,2\,3\,4 \end{pmatrix}$$

$\square$

The group S_4 is not commutative. (Find an example of permutations p, q in S_4 so that $pq \neq qp$.)

Symmetries of the square

Consider now the symmetries of a square that you worked with in the activities. Recall that we think of the square as made of rigid matter. A symmetry of the square is a transformation that repositions the square in its original place without deforming it. Two such symmetries are considered equal if their end result is the same (even if they have arrived at it via different routes).

To help you with the following discussion, you might make a cardboard square and label the four corners. It will probably be helpful to write your labels on both sides. (Actually a computer disk works pretty well). Take some large flat surface where you can trace an outline of the square. Lay the square on the outline and take note of where the vertices (corners) lie. Now think of all ways in which you can pick up the square and put it back in the same outline but with the vertices not necessarily in their original positions. Assume that you don't cut, fold or twist the sqaure, but keep it in its original rigid form. These are the symmetries of the square. What are all the possibilities for new positions?

Rotations. You could rotate the square, say in the clockwise direction. Since it must return to the original outline, you have to go at least 90 degrees. Let's call that rotation R_1. You could do that again to get a rotation of 180 degrees and yet again for a rotation of 270 degrees. Let's call these R_2, R_3 respectively. You could do it yet again, obtaining a rotation of 360 degrees. Since this gives the same end result as a "rotation" of 0 degrees we call it R_0. This is the symmetry of "doing nothing". What if you rotated the square by further steps of 90 degrees? Would that give you a symmetry? a new one? What about rotating in the counter-clockwise direction? Does that give anything new?

Flips. You could flip the square about various "axes of symmetry", so that the flipped square is back in the original position. How many such lines can you find? Figure 2.1 shows 4 lines of symmetry: horizontal, vertical and two diagonals. The corresponding symmetries are denoted by H, V, D and D'. For example, flipping in the horizontal line, interchanges the top and bottom sides of the square. Can you describe the other flips similarly? Are there any more flips?

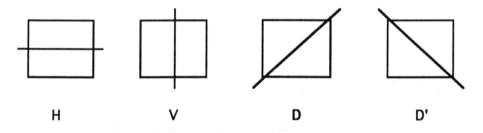

| H | V | D | D' |

FIGURE 2.1. Symmetries of the square

Combinations. We have listed a set of eight repositionings or symmetries of the square $\{R_0,\ R_1,\ R_2,\ R_3,\ H,\ V,\ D,\ D'\}$. We call that set D_4. Now what about doing one of them and then following that by another? Does that give any new ones? If you do two rotations, you just get another rotation, right? If you do a flip twice, what do you get? Now what about doing first one flip and then a different flip? Do you see that you just get the same as one of the rotations? Better use your square. Finally, what happens if you cross a rotation with a flip? Just another flip, but you might have to look hard at this one. Can you think of an easier way?

Figure 2.2 is a chart of all the symmetries you get by combining these eight. It would be helpful if you check each of the 64 entries. Most of them you should be able to do in your head, but some of them, you might have to work with your square, or with pencil and paper. In such a chart, which is sometimes called a *Cayley Table* for the group, the convention is that an element in the column at the left multiplies with an element in the row at the top to get the result at the corresponding cell in the chart. Thus, in D_4, the value of $R_3 H$ is obtained by looking at the cell on the fourth row and fifth column where is found D.

	R_0	R_1	R_2	R_3	H	V	D	D'
R_0	R_0	R_1	R_2	R_3	H	V	D	D'
R_1	R_1	R_2	R_3	R_0	D'	D	H	V
R_2	R_2	R_3	R_0	R_1	V	H	D'	D
R_3	R_3	R_0	R_1	R_2	D	D'	V	H
H	H	D	V	D'	R_0	R_2	R_1	R_3
V	V	D'	H	D	R_2	R_0	R_3	R_1
D	D	V	D'	H	R_3	R_1	R_0	R_2
D'	D'	H	D	V	R_1	R_3	R_2	R_0

FIGURE 2.2. Multiplication Table for D_4

We call D_4, together with the operation of combining two repositionings, the *Dihedral group of degree 4*. It is okay to use the word group already since you checked in the activities that the group axioms are satisfied. One last point before we write out the proof that it is a group. When you do one of these repositionings, after it is over, all that you see is the same square in the same position. The only thing that has changed are the labels of the vertices. You have the same four labels, but in some different order. You can think of this as just permuting the four vertices. Thus, combining two repositionings is just the same as taking the product of the two permutations, and we can call that *os*. This will make the checking work pretty easy.

Proposition 2.9 *The Dihedral group* $[D_4, os]$ *forms a group.*

Proof.

Closure. It can be observed from the chart that the product of two elements in D_4 is always an element of D_4. (Can you find an abstract argument?)

Associativity. Since the elements of D_4 are permutations of the vertices and the operation is permutation product, the associativity is "inherited" from S_4.

Existence of identity. As with permutations, the "do nothing" operation R_0 is the identity.

Existence of inverses. The inverse of a rotation is a rotation in the opposite direction or, what is the same thing, a rotation in the clockwise direction completing the given rotation all the way to the original position. Each flip is its own inverse. That is, if you do a flip twice, you return to the original position.

$$\square$$

The group D_4 is not Abelian as you can easily see from a glance at the chart.

What does a group chart have to look like in order for the group to be Abelian?

Groups of matrices

As a final example in this section, we consider the set M_2 of all 2×2 real matrices, that is, matrices whose entries are real numbers. We use the usual convention of writing an element of M_2 as $\begin{pmatrix} a & b \\ c & d \end{pmatrix}$ where a, b, c, d are real numbers. We define the operation in M_2 to be ordinary multiplication of matrices. That is,

$$\begin{pmatrix} a & b \\ c & d \end{pmatrix} \begin{pmatrix} e & f \\ g & h \end{pmatrix} = \begin{pmatrix} ae + bg & af + bh \\ ce + dg & cf + dh \end{pmatrix}.$$

Is M_2 with this operation a group? What would happen if you considered only *non-singular* matrices, that is those matrices $\begin{pmatrix} a & b \\ c & d \end{pmatrix}$ such that $ad - bc \neq 0$? Does this form a group?

2.1.4   Elementary properties of groups

There are several simple but useful facts that follow without too much difficulty from the definition of a group. We collect them and their proofs now in the following list of propositions. In some cases, the proof of a proposition will be left as an exercise for the reader.

Proposition 2.10 (uniqueness of identity and inverses.) *In any group, the identity element and the inverse of any element are unique.*

Proof. Let G be a group. To show the uniqueness of the identity, we suppose that e and e' are identity elements in G. We then want to show that they are the same. Consider the product $e \circ e'$. We can compute it in two ways. Since e is an identity, $e \circ e' = e'$. Since e' is an identity, $e \circ e' = e$. Hence, $e = e'$.

To show the uniqueness of the inverse, assume that a' and a'' are inverse elements for the same element $a \in G$. We then want to show that they are the same. Consider the product $a' \circ a \circ a''$. We can compute it in two ways. Since a' is an inverse for a, $(a' \circ a) \circ a'' = e \circ a'' = a''$. Since a'' is an inverse for a, $a' \circ (a \circ a'') = a' \circ e = a'$. Because of associativity, the two left hand sides are equal. Hence, $a'' = a'$.

$$\Box$$

Remark. This proposition allows us to speak of *the* identity of a group and *the* inverse of an element. In general, we denote the unique inverse of a by a^{-1}.

Product of n elements. Notice that in the previous proof we made use of the fact that, because of associativity, the triple product $a \circ b \circ c$ of three elements is unambiguously defined. In fact, a more general statement can be made. It can be proved that the product $a_1 \circ a_2 \circ \ldots \circ a_n$ of n elements in a group has the same value no matter how we parse it (that is, how we insert parentheses).

Powers in groups. In particular, the product $a \circ a \circ \ldots \circ a$ (n times) is unambiguously defined. We call this product the nth power of a and denote it by a^n. The powers of a single element satisfy the same rules of calculation as do powers in ordinary arithmetic. In analogy with arithmetic we define a^0 to be the identity element of the group and a^{-1} to be the inverse of a. Is this what you guessed in Activity 9?

Proposition 2.11 (properties of powers.) *Let a be an element of a group, a^{-1} its inverse and let n, k be positive integers.*

1. $a^{n+k} = a^n \circ a^k$

2. $a^{nk} = (a^n)^k$

3. $(a^n)^{-1} = (a^{-1})^n$

4. $a^{n-k} = a^n \circ (a^{-1})^k$

Proof. Exercise 3.

Negative powers. If n is a positive integer, we define $a^{-n} = (a^{-1})^n$. In light of part 3 of Proposition 2.11, this is also equal to $(a^n)^{-1}$.

It is not hard to show that the statements 1-4 of Proposition 2.11 hold for all integers n, k (positive, negative or 0). (Exercise 3)

In Activity 10 you studied the inverse of a product and compared it with the product of the individual inverses. Hopefully, the next proposition will not be a surprise to you.

Proposition 2.12 (inverse of product.) *The inverse of a product is the product of the inverses in reverse order. That is, let a, b be elements of a group. Then*

$$(aob)^{-1} = b^{-1} \circ a^{-1}.$$

Proof. We have to show that the product of $b^{-1} \circ a^{-1}$ and $a \circ b$ (in both orders) is equal to the identity e of the group. In fact, we have,

$$(a \circ b) \circ (b^{-1} \circ a^{-1}) = a \circ (b \circ b^{-1}) \circ a^{-1} = a \circ e \circ a^{-1} = a \circ a^{-1} = e$$

and

$$(b^{-1} \circ a^{-1}) \circ (a \circ b) = b^{-1} \circ (a^{-1} \circ a) \circ b = b^{-1} \circ e \circ b = b^{-1} \circ b = e.$$

$$[]$$

Powers in Abelian groups. If the group is commutative, then we have the relation

$$(a \circ b)^n = a^n \circ b^n$$

where a, b are elements of the group and n is an integer (positive, negative or 0). This relation does not hold in non-commutative groups. Compare this with your findings in Activity 11.

Left identity and left inverses suffice. Consider the definition of the identity element e in a group G: for all $a \in G$, $e \circ a = a \circ e = a$. Because e appears on both sides of a in this condition, it is sometimes called a *two-sided* identity. If e is only assumed to satisfy: for all $a \in G$, $e \circ a = a$, then it is called a *left* identity. *Right* identity is defined in the same way.

Similarly, *left, right* and *two-sided* inverses are defined.

In the examples above, we always proved the existence of a two-sided identity and two-sided inverses. Actually, it is enough to show the existence of a left identity and a left inverse for every element. That is, if we have a set together with a binary operation which is closed, associative, has a left identity and has a left inverse for every element, then it can be shown that it is a group. On the other hand left identity and right inverses are not sufficient. (Exercise 4.)

Multiplicative and additive notation. The operation $\circ$ in a group is often thought of as analogous to multiplication. As is usually done with multiplication, we will often drop it completely and write ab for the product $a \circ b$.

Sometimes however, especially when working with number systems as examples of groups, it is customary to use additive notation, that is, $a + b$ for $a \circ b$. In this case, we use 0 for the identity, $-a$ for the inverse of a and na for the nth power of a.

2.1.5 EXERCISES

1. Show that commutativity of $[\mathcal{Z}_{12}, +_{12}]$ is inherited from ordinary addition.

2. We have seen that in $[\mathcal{Z}_{12} - \{0\}, *_{12}]$, we have $2 *_{12} 6 = 0$. Deduce from this relation that 2 and 6 are not invertible in this system.

3. Prove Proposition 2.11 in its original form and in its stronger form, that is, when n, k are taken to range over *all* integers (positive, negative and 0).

4. Show that a set together with a binary operation that is closed, associative, has a left identity and that has a left inverse for every element, is a group. Give an example to show that a set together with a binary operation that is closed, associative, has a left identity and that has a *right* inverse for every element is not necessarily a group.

5. Assuming that $[\mathcal{Z}, +]$ is a group and using the definition of $\mathcal{Q}$ as the set of all numbers that can be written in the form a/b, $a, b \in \mathcal{Z}$, $b \neq 0$, show that $[\mathcal{Q}, +]$ is a group.

6. Show that the set M_2 of all 2×2 real matrices with the operation of matrix multiplication (see p. 52) is not a group, even if the zero matrix is removed.

7. Determine if the set of all non-singular matrices in M_2 (see p. 52) is a group.

8. Show that two integers a and b have the same remainder upon division by n if and only if their difference $a - b$ is exactly divisible by n.

   (When this relation holds we write $a \equiv b \pmod{n}$ and read this as a *is congruent to b mod n*.)

9. Solve the following congruences (i.e. find the set of all solutions; in case there are no solutions, this will be the empty set):

$$x \equiv 1 \pmod{5}$$
$$x \equiv 2 \pmod{5}$$
$$2x \equiv 3 \pmod{9}$$
$$2x \equiv 3 \pmod{6}$$
$$ax \equiv b \pmod{n}$$

10. Show that the relation $\equiv \pmod{n}$ is an equivalence relation on $\mathcal{Z}$; that is, it satisfies the following three properties:

Reflexivity: For all $a \in \mathcal{Z}, a \equiv a \pmod{n}$

Symmetry: For all $a, b \in \mathcal{Z}, a \equiv b \pmod{n} \Rightarrow b \equiv a \pmod{n}$

Transitivity: For all $a, b, c \in \mathcal{Z}, a \equiv b \pmod{n}$ and $b \equiv c \pmod{n} \Rightarrow a \equiv c \pmod{n}$

11. Characterize the equivalence class $[a]$ of $a \in \mathcal{Z}$ relative to the relation $\equiv \pmod{n}$, that is, the set $\{b \in \mathcal{Z} : a \equiv b \pmod{n}\}$.

12. Give an abstract proof of the fact that $[D_4, os]$ is closed (c.f. Proposition 2.9, p. 51).

13. Write the multiplication table of D_3, the dihedral group of degree 3 (that is the group of all symmetries of an equilateral triangle). How many elements are there? Is it Abelian?

14. Let a, b be two commuting elements in a group (i.e., $ab = ba$), and let n be a positive integer. Show that $(ab)^n = a^n b^n$. Show that this holds for any integer n.

15. Show that each one of the following conditions on a group $[G, o]$ implies that the group is Abelian.

    For all $a, b \in G, (ab)^{-1} = a^{-1} b^{-1}$.

    For all $a, b \in G, (ab)^2 = a^2 b^2$.

    For all $a \in G, a^2 = e$.

16. Show that every group of order 4 (that is, has four elements) or less is Abelian.

17. If G is a group of even order, show that there must be an element $a \neq e$ satisfying $a^2 = e$.

18. Find all the elements of S_3 satisfying $x^2 = e$ and all the elements satisfying $x^3 = e$.

19. Do the same as the previous exercise for S_4.

20. Find three pairs of elements a, b of S_5 for which $ab = ba$ and three pairs for which $ab \neq ba$.

21. Consider the permutation $p = \begin{pmatrix} 1\,2\,3\,4\,5\,6 \\ 2\,3\,4\,5\,6\,1 \end{pmatrix} \in S_6$. Find n such that $p^n = e$, the identity permutation.

22. Find a two-element subset of S_3 which in itself forms a group (relative to permutation product). What about a three element set? A four element set?

23. Suppose G is a finite group, $a \in G$. Show that there is a positive integer n such that $a^n = e$. Deduce that in a finite group the inverse of an element can always be expressed as a power of that element.

24. Look at Propositions 2.6 and 2.7 (p. 46). State and prove a general result of which these two propositions are special cases.

2.2 The modular groups and the symmetric groups

2.2.1 ACTIVITIES

1. You have worked with Z5, Z6, Z12 and addition mod the appropriate integer. It should be clear to you by now that this can be done for any positive integer. Write two funcs Zmod and amod. Both of these funcs accept a positive integer. The func Zmod returns the set of integers mod$n$. The func amod returns a func which implements addition mod n.

    Hint. Look back in Chapter 1 to see examples of a func returning a func. See Activities 10 and 11 in Section 1.3.1, p. 30.

2. Write a func that accepts a pair [a,b] of non-negative integers and if b>0 returns the pair [b, a mod b]. For this activity you are to apply your func to a pair [a,b], apply it again to the resulting pair and continue to do that as long as it makes sense. Do this to the following pairs.

$$[704990, 3698331]$$

$$[12966835, 572286]$$

    Can you figure out what is going on? Try it on some pairs of small numbers and make guesses.

    Division-with-remainder. Division with remainder is taught in schools from an early age. Thus when 37 is divided by 12 the result is 3 with a remainder of 1. There are several ways to write this fact, e.g. $37/12 = 3(1)$, or more accurately, $37/12 = 3 + 1/12$. For our purposes, and in number theory in general, it is convenient to write this last relation using only integers: $37 = 3 * 12 + 1$. (Note that the remainder must always be less than the divisor.) The general form of division by 12 is given by the formula

$$a = 12q + r, \quad 0 \le r < 12,$$

    where a is any integer, q is the "quotient" and r is the "remainder". The case $r = 0$ arises when 12 exactly divides a, e.g. when $a = 36$. We then say that a is a multiple of 12, or that 12 divides a.

3. For each of the pairs in Activity 2, write a succession of division formulas of the form $a = bq + r$ representing the calculations that your **func** made. You might want to write another **func** that gives you q as well. In fact, the quotient is given by a **div** b, so the full formula is

$$a = b(a \text{ div } b) + (a \text{ mod } b).$$

4. Relations like $x \bmod 5 = y \bmod 5$ will be very important in your study of groups. Write out as many different ways as you can to test two integers x, y to see if this relation holds. Explain each of your ways.

5. How many permutations are there in S_3? S_4? S_5? S_6? S_{10}? You might want to use the **ISETL** operation # for some of these.

6. The support of a permutation is the set of numbers which are actually changed. Thus, the support of the permutation $\begin{pmatrix} 1\,2\,3\,4\,5\,6\,7\,8 \\ 5\,6\,3\,4\,1\,8\,7\,2 \end{pmatrix}$ is $\{1, 2, 5, 6, 8\}$. Write an **ISETL func support** that will accept a permutation and return its support. Check your func on this permutation, and also the following four. $\begin{pmatrix} 1\,2\,3\,4\,5 \\ 4\,5\,1\,2\,3 \end{pmatrix}$,

$$\begin{pmatrix} 1\,2\,3\,4\,5\,6\,7\,8\,9\,10\,11 \\ 1\,6\,4\,7\,5\,2\,3\,8\,11\,10\,9 \end{pmatrix}, \begin{pmatrix} 1\,2\,3\,4\,5\,6\,7\,8 \\ 2\,3\,4\,5\,6\,7\,8\,1 \end{pmatrix}, \begin{pmatrix} 1\,2\,3\,4\,5\,6\,7\,8 \\ 1\,2\,4\,7\,5\,6\,3\,8 \end{pmatrix}.$$

7. Write an **ISETL func is_disjoint** that will accept two permutations and check if their supports are disjoint. (Recall that two sets are disjoint if their intersection is empty.) Check your func on all pairs of permutations you can get from the previous activity.

8. Consider the permutation $\begin{pmatrix} 1\,2\,3\,4\,5\,6\,7\,8 \\ 5\,6\,8\,2\,1\,4\,7\,3 \end{pmatrix}$. Pick any one of the numbers $\{1, 2, ..., 8\}$ being permuted and apply the permutation to that number, then to the result and so on. Write down on one line all of the numbers that you get. Now pick one of the remaining numbers that did not appear and do the same thing, writing the numbers that you get on the next line. Continue this until all of the numbers $\{1, 2..., 8\}$ have been used up.

What do you see in the results of this activity? Write down all of the observations you can make about what came out.

9. In the previous Activity you took a permutation p and a number a and wrote a list of all the numbers you get by applying p successively.

This list is called the *orbit* of a under *p*. Here is an **ISETL** func that will compute orbits for you. Note that the interpretation of the tuple that orbit returns is different from the interpretation of the tuple that orbit accepts. Describe this difference.

```
orbit := func(a,p);
            local a0, orb;
            a0 := a; orb := [a0];
            while p(a) /= a0 do
               orb := orb with p(a);
               a := p(a);
            end;
            return orb;
         end;
```

Write out in words, as carefully as you can, the process that orbit goes through in calculating an orbit.

Apply orbit to find all of the orbits of the permutations in Activity 7.

10. Here is an **ISETL** func that will accept a permutation and return the set of all its orbits. It uses the func support which you wrote as part of Activity 6 and the func orbit of Activity 9. Actually, because we are using the support, our func will ignore the trivial orbits, i.e. those with only one element.

```
pdec := func(p);
            local a, s, orbs;
            s := support(p);
            orbs := [];
            while s /= {} do
               a := arb(s);
               orbs := orbs with orbit(a,p);
               s := s - {x: x in orbit(a,p)};
            end;
            return orbs;
         end;
```

Write out an explanation of how this pdec works.

11. It can happen that a permutation has only one non-trivial orbit. That is, if you take a number, and apply the permutation successively, you get all of the numbers in the support of the permutation. Such a permutation is called a *cycle*. The permutation $\begin{pmatrix} 1\,2\,3\,4\,5\,6 \\ 4\,2\,6\,3\,5\,1 \end{pmatrix}$ is a cycle. Notice that we can write such a cycle in the following more

compact form, called cycle notation: (3 6 1 4). We read this as "3 goes to 6, 6 goes to 1, 1 goes to 4, and 4 goes to 3". Since 2 and 5 are not changed, they do not need to be mentioned. All this can be abbreviated as $3 \rightarrow 6 \rightarrow 1 \rightarrow 4 \rightarrow 3$. Does the cycle (1 4 3 6) represent the same permutation or a different one? (For technical reasons we choose not to call the identity permutation a cycle although it can still be written in cycle notation as $e = (1) = (2) = \cdots$.) Write a func is_cycle to test if a given permutation is a cycle. Use your func to find all cycles in S_3 and S_4.

12. For each of the following permutations in S_4, find all its orbits. Which of the permutations are cycles?

$$\begin{pmatrix} 1\,2\,3\,4 \\ 2\,4\,1\,3 \end{pmatrix},$$

$$\begin{pmatrix} 1\,2\,3\,4 \\ 3\,2\,1\,4 \end{pmatrix},$$

$$\begin{pmatrix} 1\,2\,3\,4 \\ 2\,1\,3\,4 \end{pmatrix},$$

$$\begin{pmatrix} 1\,2\,3\,4 \\ 2\,3\,4\,1 \end{pmatrix},$$

$$\begin{pmatrix} 1\,2\,3\,4 \\ 4\,3\,2\,1 \end{pmatrix},$$

$$\begin{pmatrix} 1\,2\,3\,4 \\ 2\,1\,4\,3 \end{pmatrix},$$

$$\begin{pmatrix} 1\,2\,3\,4 \\ 1\,4\,3\,2 \end{pmatrix}.$$

13. Since cycles are permutations, we can form the product of two cycles p and q. We write pq which means apply the cycle q and then, to the result of that, apply the cycle p. Suppose that the two cycles have disjoint supports. (We abbreviate this by calling them *disjoint cycles*.) What can you say about their product? Look at examples taken from the cycles you produced in Activity 12. Write pairs of them in cycle notation and compute their products. See what general relations you can surmise.

2.2.2   The modular groups $\mathcal{Z}_n$

In the activities preceding this and the previous section you worked with various systems of integers mod n. Then in the previous section the mathematical contents of these activities were summarized. In particular, some systems involving integers mod n were shown to be groups; others were shown not to be. We are now going to introduce these systems more systematically and develop some tools that, in the end, will enable us to prove or disprove all group properties for the various systems of integers mod n, for all n.

Inheritance. A powerful method that will be used in developing the properties of the integers mod n has many applications in other situations as well. In this method, we show how integers mod n "inherit" many of their properties from the correponding properties of the usual integers. (We took a first look at inheritance when discussing associativity of $\mathcal{Z}_{12}$ in the proof of Propostion 2.4, p. 45.) To do this, we will establish an intimate connection between the regular integers and the integers mod n. In the following discussion we assume that an arbitrary but fixed integer $n > 1$ is given (for specific examples we will mostly use $n = 5$).

In the activities you saw different ways of expressing the division with remainder of two integers. Thus, the fact that when 11 is divided by 5 the result is 2 with a remainder of 1 can be written as $11 = 2*5+1$. In general, if a divided by b gives quotient q and remainder r, this can be expressed in the form $a = qb + r$. Note that the remainder must always be non-negative and strictly less than the divisor. The fact that such division can always be done is an important property of the integers, which we give as a theorem.

Theorem 2.1 (division with remainder.) *Given $a, b \in \mathcal{Z}$, $b \neq 0$, there exist $q, r \in \mathcal{Z}$ such that $a = qb + r$ and $0 \leq r < |b|$.*

Proof. First assume that $a \geq 0$, $b > 0$. If $b > a$ we take $q = 0$, $r = a$. If $b \leq a$ then let q be the largest integer such that $qb \leq a$ and set $r = a - bq$. It is easy to check that q, r have the desired properties.

The case of one or both of a, b being negative is left as an exercise (Exercise 1).

$\square$

Definition 2.2 *In the context of the previous theorem, in case $r = 0$, we say that b divides a, and write $b|a$.*

Our next step is to notice that the **ISETL** object a mod n that we have been working with is nothing else than the remainder we get when dividing a by n. For example, since 11 = 2*5 + 1, it follows that 11 mod 5 = 1. In general, if we express the division of a by n in the form

$$a = qn + r, \quad 0 \leq r < n,$$

r is just a mod n. Sometimes, when there is no danger of confusion, we denote a mod n by $\bar{a}$ (read "a-bar") for short.

For example, if $n = 5$ then $\bar{7} = 2$, $\overline{23} = 3$ and $\bar{4} = 4$. Also, $\bar{5} = \overline{10} = \overline{15} = 0$ and $\bar{6} = \overline{11} = \overline{16} = 1$.

Definition 2.3 (congruence mod n.) *Integers $a, b \in \mathcal{Z}$ are called con-gruent (mod n), denoted $a \equiv b$ (mod n), if they have the same remainder mod n, that is, $a \bmod n = b \bmod n$.*

For example,

$$5 \equiv 10 \equiv 15 \equiv 0 \pmod 5$$

$$6 \equiv 11 \equiv 16 \equiv 1 \pmod 5.$$

In this example, numbers that are congruent mod 5 differ by multiples of 5. Another way of saying this is that their difference is divisible by 5. This is generalized in the next proposition, which also relates to your investigations in Activity 5.

Proposition 2.13 (criteria for congruence mod n.) *For $a, b \in \mathcal{Z}$, the following conditions are equivalent:*

1. $a \equiv b \pmod n$.

2. $a = b + kn$ for some $k \in \mathcal{Z}$.

3. $n|(a - b)$ *(recall that this is read "n divides $a - b$" and it means that $a - b$ is a multiple of n).*

Proof. Exercise 5.

Recall that doing addition mod 12 of two numbers means first finding their usual sum as integers, then taking the remainder of the sum upon division by 12. This is formalized and generalized in the following definition.

Definition 2.4 (addition and multiplication mod n.) *For $a, b \in \mathcal{Z}$ we define*

$$a +_n b = (a + b) \bmod n$$
$$a *_n b = (a * b) \bmod n.$$

For example, $3 +_5 4 = 2$, $2 +_5 2 = 4$, $2 +_5 3 = 0$, $101 +_5 57 = 3$, $1 *_5 3 = 3$, $2 *_5 3 = 1$, $7 *_5 9 = 3$, $2 *_6 3 = 0$.

Definition 2.5 (integers mod n.) *We define $\mathcal{Z}_n = \{0, 1, ..., n - 1\}$ and call the elements of $\mathcal{Z}_n$ integers mod n . (The name is explained by noting that $\mathcal{Z}_n = \{a \bmod n : a \in \mathcal{Z}\}$.)*

Remark. $\mathcal{Z}_n$ is closed with respect to addition and multiplication mod n. This follows directly from the definitions.

We can now establish the "intimate connection" between the integers and the integers mod n, which was mentioned in the remark on inheritance.

First we note that this connection, basically relating regular addition and multiplication of integers with these operations mod n, already appeared in the activities preceding this section. For example, you saw that if $a = 17$, $b = 9$ and $n = 6$ then

$$(a + b) \bmod n = (a \bmod n) +_n (b \bmod n),$$

or using the present notation,

$$\overline{a + b} = \bar{a} +_n \bar{b}.$$

In other words, if you want to calculate $\overline{a + b}$, you can either calculate $a + b$ and then find the remainder after division by 6, or you can find the remainders of both a and b on division by 6, add them and then take the remainder on division by 6.

Theorem 2.2 (the fundamental connection between $\mathcal{Z}$ and $\mathcal{Z}_n$.) *Let $a, b \in \mathcal{Z}$ and let $\bar{t}$ denote $t \bmod n$ for some fixed n. Then*

1. $\overline{a + b} = \bar{a} +_n \bar{b}$

2. $\overline{a * b} = \bar{a} *_n \bar{b}$

Proof. We will prove 1, leaving 2 as an exercise. The idea is to calculate the remainder of $a + b$ upon division by n, using the individual remainders of a and b. To express the remainders of a and b, we write out their division by n:

$$a = qn + r, \quad 0 \le r < n$$
$$b = pn + s, \quad 0 \le s < n.$$

Thus $\bar{a} = r$ and $\bar{b} = s$. Also,

$$a + b = (qn + r) + (pn + s) = (q + p)n + (r + s).$$

Since this shows that $a + b$ and $r + s$ differ by a multiple of n, it follows from the criteria for congruence that

$$\overline{a + b} = \overline{r + s}.$$

On the other hand,

$$\overline{r + s} = \overline{\bar{a} + \bar{b}} = \bar{a} +_n \bar{b} \quad \text{(by the definition of addition mod} n\text{.)}$$

Thus $\overline{a + b} = \bar{a} +_n \bar{b}$.

$$\square$$

Now we have the tools needed to settle the question of which of the various systems of integers mod n is or is not a group.

Proposition 2.14 *The integers mod n with addition mod n form an Abelian group.*

Proof.

Closure. This has already been noted following the definition of $\mathcal{Z}_n$.

Associativity. We will show how this property is "inherited" from the associativity of $\mathcal{Z}$, via the fundamental connection. Indeed, for all $a, b, c \in \mathcal{Z}$,

$$(a + b) + c = a + (b + c) \qquad \text{(associativity in } [\mathcal{Z}, +] \text{)}$$

$$\overline{(a + b) + c} = \overline{a + (b + c)} \qquad \text{(remainders of equal numbers are equal)}$$

$$\overline{a + b} +_n \overline{c} = \overline{a} +_n \overline{b + c} \qquad \text{(the fundamental connection)}$$

$$(\overline{a} +_n \overline{b}) +_n \overline{c} = \overline{a} +_n (\overline{b} +_n \overline{c}) \qquad \text{(the fundamental connection once again)}$$

If, in particular, we have $a, b, c \in \mathcal{Z}_n$, then $\overline{a} = a$, $\overline{b} = b$, $\overline{c} = c$ and the last equality is the required associative property of $\mathcal{Z}_n$.

Commutativity. This again is done by inheritance. (Exercise 8)

Existence of identity. 0 is an identity, for if $a \in \mathcal{Z}_n$ then

$$a +_n 0 = \overline{a} +_n \overline{0} = \overline{a + 0} = \overline{a} = a.$$

(Note that this property of 0 is once again inherited from the same property of the regular addition in $\mathcal{Z}$.)

Existence of inverses. If $a \in \mathcal{Z}_n$ then $n - a$ is also in $\mathcal{Z}_n$ (why?) and is the inverse of a, since

$$a +_n (n - a) = \overline{a + n - a} = \overline{n} = 0$$

$\Box$

Having established that $\mathcal{Z}_n$ forms a group with respect to addition mod n, it is now natural to inquire about multiplication mod n. Of course $[\mathcal{Z}_n, *_n]$ cannot be a group since, as we know, 0 is not invertible. We therefore ask, is $\mathcal{Z}_n - \{0\}$ a group with respect to multiplication mod n?

As we have seen in the activities and in the previous section, the answer is "sometimes yes, sometimes no". For example, $\mathcal{Z}_6 - \{0\}$ is not a group with respect to $*_6$, but $\mathcal{Z}_5 - \{0\}$ is a (commutative) group with respect to $*_5$. We wish to determine just what it is that makes these systems a group for some values of n and not a group for others. We might say right away that associativity, commutativity and existence of identity (1 is the identity element) are inherited from $[\mathcal{Z}, *]$ the same way they were for addition. Also, closure of $\mathcal{Z}_n$ (but not necessarily of $\mathcal{Z}_n - \{0\}$) follows immediately from the definitions. The only remaining questions are therefore whether the product mod n of two non-zero elements of $\mathcal{Z}_n$ is again non-zero, and the existence of inverses.

We have seen for example that $2 *_6 3 = 0$ and therefore $\mathcal{Z}_6 - \{0\}$ is not closed. In fact, the same argument works for all non-prime n: If $n = mk$, where $1 < m, k < n$ then m, k are in $\mathcal{Z}_n - \{0\}$ but their product mod n (which is 0) is not. Thus if n is not prime, $\mathcal{Z}_n - \{0\}$ is not a group. The converse of this observation is an important fact of elementary number theory. Both are summarized in the following theorem.

Theorem 2.3 (multiplicative groups mod n.) *For a positive integer n, $\mathcal{Z}_n - \{0\}$ forms a group with respect to multiplication mod n if and only if n is prime.*

Preliminary discussion of the proof. In view of the discusssion preceding the theorem, we need prove one direction only, that is, if p is prime then $\mathcal{Z}_p - \{0\}$ is a group. The closure issue is relatively easily settled. We need to show that if a and b are non-zero elements in $\mathcal{Z}_p$ then $a *_p b \neq 0$. Can you see how this follows from the following fundamental property of any prime number p?

> If a, b are integers and if p divides the (ordinary) product ab then p must divide a or b.

As for the proof of the existence of inverses, you'll have the chance to discover it for yourself in the exercises.

2.2.3   The symmetric groups S_n

In the activities and in the previous section you investigated some sets of permutations with respect to the operation of permutation product. In particular, you looked at the set S_3 of all permutations of $\{1, 2, 3\}$ and at the set S_4 of all permutations of $\{1, 2, 3, 4\}$, and found that both form groups with respect to the operation of permutation product. In this section we are going to generalize these findings. We will define S_n for any natural number n, generalize the definition of permutation product to S_n, and prove that S_n forms a group with respect to this product.

We start by giving a general definition of permutations. Recall that we started with a permutation as reordering, represented by a `tuple`, then moved to the two-row standard notation and to viewing permutations as functions. (See Section 1.2.6, p. 17 and Section 2.1.3, p. 47ff.) In terms of functions, the condition that all numbers in the bottom row should be different means that the function is one-to-one (1-1) and the condition that all the numbers appear in the bottom row mean that the function is onto. Recall that a function $f : A \rightarrow B$ is said to be one-to-one if for all $a, a' \in A$, $a \neq a'$ implies $f(a) \neq f(a')$. (This is often put in the equivalent form: for all $a, a' \in A$, $f(a) = f(a')$ implies $a = a'$.) The function f is said to be onto B if every b in B has a "source", that is, there exists $a \in A$ with $f(a) = b$.

Definition 2.6 (permutations.) *A permutation of the set $\{1, 2, ..., n\}$ is a one-to-one function $p : \{1, 2, ..., n\} \rightarrow \{1, 2, ..., n\}$. If $p(i) = a_i$, $i = 1, ..., n$, then the permutation p is usually written as*

$$\begin{pmatrix} 1 & 2 & \cdots & n \\ a_1 & a_2 & \cdots & a_n \end{pmatrix}.$$

The set of all permutations of $\{1, 2, ..., n\}$ is denoted S_n.

Remark. Because we are only dealing with permutations of a finite set, "one-to-one" implies "onto".

Indeed, consider a permutation p in S_n, given in its standard notation

$$\begin{pmatrix} 1 & 2 & \cdots & n \\ a_1 & a_2 & \cdots & a_n \end{pmatrix}.$$

The assumption that p is 1-1 means that the numbers in the bottom row are all different. Hence there must be n different numbers in the bottom row, all taken from $\{1, ..., n\}$. Thus each of the numbers $1, ..., n$ must appear somewhere in the bottom row. This means that the function p is onto.

The converse of this remark is also true (Exercise 2).

The number of elements of S_n. How many permutations are there in S_n? Let us count the ways to fill in the bottom row in the standard notation

$$\begin{pmatrix} 1 & 2 & \cdots & n \\ a_1 & a_2 & \cdots & a_n \end{pmatrix}.$$

There are n different choices for a_1. Once we have chosen a value for a_1, there are $n-1$ choices for a_2, namely, all the numbers except the one chosen for a_1. For each choice of values for a_1 and a_2, there are $n - 2$ choices for a_3, etc. Finally, when all but the last element have been assigned a value, there remains only one choice for a_n. The number of choices for the bottom row is therefore $n!$, the product of all integers from 1 to n.

For example, $\#S_2 = 2$, $\#S_3 = 6$, $\#S_4 = 24$.

Definition 2.7 (product of permutations.) *The product pq of permutations p, q in S_n is defined by*

$$(pq)(i) = p(q(i)), \quad i = 1, ..., n.$$

That is, to evaluate pq on a certain number, apply first q to that number, then apply p to the result. Note that though p is the first factor in pq, the definition specifies an inversion: first apply q, then p.

We have seen in the previous section that S_3 and S_4 are groups. We now generalize this for every n.

Proposition 2.15 S_n *forms a group with respect to product of permutations.*

This group is called the *symmetric group* of degree n.

Proof.

Since permutations and their product are now defined using standard terminology of functions, validating the group axioms becomes largely a matter of citing the relevant facts about functions.

Closure. The composition of two 1-1 functions of $\{1, 2, ..., n\}$ onto itself is again a 1-1 function of $\{1, 2, ..., n\}$ onto itself (Exercise 3). Thus the product of two permutations in S_n is again a permutation in S_n.

Associativity. This is a property of the composition of functions in general (Exercise 11).

Existence of identity. The so-called "identity function" I, defined by

$$I(k) = k \quad \text{for all } k \in \{1, 2, ..., n\},$$

is the identity element. Indeed, if $p \in S_n$ then

$$(Ip)(k) = I(p(k)) = p(k) \text{ for all } k \in \{1, 2, ..., n\}$$

hence $Ip = p$.

Note that in the standard notation I is written as two equal rows:

$$\begin{pmatrix} 1 \; 2 \; ... \; n \\ 1 \; 2 \; ... \; n \end{pmatrix}.$$

Existence of inverses. If a function $f : A \to A$ is 1-1 and onto then it has an inverse function $g : A \to A$ which is also 1-1 and onto (Exercise 11).

$\Box$

Note that if $p \in S_n$ is given by

$$\begin{pmatrix} 1 & 2 & ... & n \\ a_1 & a_2 & ... & a_n \end{pmatrix},$$

then the inverse of p is given by

$$\begin{pmatrix} a_1 & a_2 & \ldots & a_n \\ 1 & 2 & \ldots & n \end{pmatrix}.$$

Remark. For $n > 2$, S_n is not commutative (why?).

Orbits and cycles

We now summarize and formalize some of the information you obtained in Activities 7-15 of Section 2.2.1, p. 58ff. First the support which is the set of digits that a permutation actually moves.

Definition 2.8 (Support of a permutation.) *The* support *of a permutation p of $\{1, 2, ..., n\}$ is given by*

$$\text{supp}(p) = \{i \in \{1, 2, .., n\} \mid p(i) \neq i\}.$$

Thinking about a permutation as a function, it should be clear that you know everything you always wanted to know about it, once you know what it does to each element in its support. In connection with supports, you saw several examples of pairs of permutations whose supports are disjoint. Did you notice that the product of such pairs always commutes? It does.

Proposition 2.16 *If p, q are two permutations in S_n whose supports are disjoint, then they commute, that is, $pq = qp$.*

Proof. Exercise 13.

In working with orbits in the activities you applied a permutation p successively to a number j. In all of the examples you worked with, did you notice that this sequence never failed to return to j and, moreover, it had no other repetitions before returning to j? This was no accident (see the exercises at the end of this section).

In the activities you saw that the orbit of a number j under a permutation p is the sequence of numbers you get by successively applying p to j. This is formalized in the following definition.

Definition 2.9 (orbits.) *Let $p \in S_n$ and $i \in \{1, 2, ..., n\}$. Assume that p is not the identity permutation. The orbit of i under p is the sequence $i, p(i), p^2(i), ..., p^{r-1}(i)$ where r is the smallest positive integer for which $p^r(i) = i$.*

Recall that in the activities you took a permutation and computed its orbits. A permutation which has only one orbit is called a *cycle*. This means that the orbit of any number in the support of a cycle is equal to the entire support.

It follows from the minimality of r (in the definition of orbits) that the integers in an orbit are all distinct. There are a few other properties of cycles that were exemplified in the activities. They are discussed in the exercises.

Recall the convenient *cycle notation* introduced in the activities. We simply list the orbit. The action of the permutation is the passage from a number to the next number, with the last number going (cycling back) to the first. Any number that does not appear on the list is not in the support of the permutation and so is not affected by its action. Thus, cycle notation for the permutation

$$p = \left(\begin{array}{ccccc} 1 & 2 & 3 & 4 & 5 \\ 5 & 1 & 2 & 4 & 3 \end{array} \right)$$

could be (1 5 3 2). It could also be (5 3 2 1) or anything you get by "cycling" the list without changing its order. Notice that the absence of 4 means that 4 is not in the support of this permutation, that is, $p(4) = 4$. There is one special case. We choose not to call the identity permutation a cycle, but it can still be represented in cycle notation as (1).

The support of a cycle is easy to determine. Do you see how you can just read it off from the cycle notation?

2.2.4 EXERCISES

1. Complete the proof of Theorem 2.1, p. 61. That is, show that division with remainder is still possible if one or both of a, b are negative.

2. Consider a finite set A and a function $f : A \rightarrow A$. Show that f is one-to-one if and only if it is onto.

3. Prove that the composition of two one-to-one functions is one-to-one.

4. Prove or give a counter-example to the assertion that the composition of two onto functions is onto.

5. Prove Proposition 2.13 on congruence mod n.

6. Prove Theorem 2.2 part 2:

$$\overline{a * b} = \overline{a} *_n \overline{b}$$

   where $\overline{x}$ means x mod n.

7. Calculate $(17 + 9) \bmod 6$ in two ways. Now consider $(ab) \bmod 25$ where $a = 342563228$ and $b = 1455567832$. One of the two ways is a little easier. Use that and the fact that the remainder on division by 25 depends only the last two digits (why?) to calculate $ab \bmod 25$ in your head.

8. Use inheritance to prove commutativity of $+_n$ in Proposition 2.14.

9. Complete the proof of Theorem 2.3; that is, show that if p is prime then $G = \mathbb{Z}_p - \{0\}$ is a group under the operation $*_p$.

10. For each positive integer a, let $[a]$ be the congruence class of a mod 3 (that is, the set consisting of all integers which are congruent to a mod 3). Describe the class $[1]$. How many different classes do we get when a ranges over all the integers? Define a binary operation on these congruence classes which will make them into a group. What familiar group does this new group resemble?

11. Complete the proof that S_n is a group (Proposition 2.15).

12. Show that for $n > 2$, S_n is not commutative.

13. Prove Proposition 2.16: If p, q are two permutations in S_n whose supports are disjoint, then they commute, that is, $pq = qp$.

14. How many different 2-cycles are there in S_3? How many 3-cycles?

15. How many different 2-cycles, 3-cycles and 4-cycles are there in S_4?

16. If $k \leq n$, how many different k-cycles are there in S_n?

17. What is the inverse of the cycle $(1\ 5\ 3\ 2)$? Of $(a_1\ a_2\ a_3 \ldots a_n)$?

18. Show that $(1\ 4\ 3\ 2)(2\ 5\ 7\ 6) = (1\ 4\ 3\ 2\ 5\ 7\ 6)$.

    Is it true that the product of two cycles having exactly one digit in common is always a cycle?

19. If p is a k-cycle, show that k is the smallest positive integer such that $p^k = I$, the identity permutation.

20. Show that if a permutation p can be written as a product of disjoint 2-cycles then $p^2 = I$.

21. Suppose that a permutation $p \in S_{10}$ has been written as a product of disjoint 2-cycles and 3-cycles. What would be the smallest positive integer n such that $p^n = I$? Same question for a product of disjoint 4-cycles and 6-cycles.

22. Prove that the permutation $(1\ 2\ 3\ 4)(5\ 6)$ cannot be expressed as a product of two disjoint 3-cycles or three disjoint 2-cycles.

23. Express the permutation $(1\ 2\ 3)$ as a product of 2 (not necessarily disjoint) 2-cycles in S_3. Prove that $(1\ 2\ 3)$ *can not* be expressed as a product of *three* 2-cycles.

24. Explain how you can read off the support of a permutation which is a cycle from its representation in cycle notation.

2.3   Properties of groups

2.3.1   ACTIVITIES

1. It is helpful for everyone to have a uniform notation when working
with a mathematical system. Therefore, we suggest you place the
following proc in a file so that it will be available to you. The funcs
identity and inverse were described in Chapter 1, p. 29.

```
name_group := proc(set,operation);
              G := set; o := operation;
              e := identity(G,o);
              i := |g -> inverse(G,o,g)|;
              writeln "Group objects defined: G, o, e,
              i.";
          end;
```

Run name_group on the following pairs of a set and an operation.

(a) Z_{12}, a_{12}.

(b) Z_5, a_5.

(c) Z_6, a_6.

(d) $Z_6 - \{0\}, m_6$.

(e) Z_5, m_5.

(f) S_3, os.

Any time you intend to work with a group we suggest that you first
apply name_group to it and then work exclusively with the standard
notation G, g .o h, e, and i(g).

2. Run the following **ISETL** code.

```
name_group(Z12, a12);
a := 3; b := 5; c := 7;
a /= b; c .o a = c .o b;
```

```
name_group(S3, os);
a := [1,3,2]; b := [3, 1, 2] ; c := [1,2,3];
a /= b; c .o a = c .o b;
```

Find an example of a binary system (a set with a binary operation)
with three elements a,b,c such that a /= b but c .o a = c .o b.

In some binary systems you cannot find such an example. That is, the
system has the property that if you take two elements and multiply

them on the left by the same element, then the two results can only be the same if the original two elements were the same. This property is called the *left cancellation law*. Similary, there is a *right cancellation law*.

Write an **ISETL** func that accepts a binary system and tests whether or not it satisfies the left cancellation law.

3. What is the relation between the left cancellation law and the property of being a group? Use your func from the previous activity and the examples of binary operations you have been working with to investigate this question. Make a general statement about this relation that "explains" (that is, is consistent with) all of the facts that you observe.

4. How many groups are there with just three elements? How many ways are there of filling in the following table so that it represents a group?

	e	a	b
e			
a			
b			

What relationship is there between the two questions we just asked?

5. Consider the group $\mathcal{Z}$ with ordinary addition. Is the subset $\{0, 1, 2, 3, 4\}$ a group? What about the subset of all even integers? All odd integers?

When is a subset of a group a group? Look over all of your examples of groups and pick out subsets. Find 5 subsets (other than the group itself) which are groups and 5 non-empty subsets which are not groups.

2.3.2 The specific and the general

In this chapter, our main emphasis has been on examples of groups. In the first section, you had an opportunity to work with specific groups and gain experience on which the more theoretical discusssion of examples in the second section was based. The idea is that before you can think about "groups in general", you must be on familiar terms with a number of specific groups.

You should be at that point now, so we can turn in this section to some general investigations of groups. Our abstract groups will never stray very far from the important examples, and the relationships we discover will also be clearly realized in one or more of the specific groups you are familiar with.

We begin with some comments on the relationship between the specific and the general. We will see how this looks in the case of the cancellation law. Then we will look into the number of elements (the order) of a group and see how much freedom there is in constructing a group, once you have decided how many elements it is to have. Finally, we end this chapter with a bridge to the next one by seeing how the powers of a single element lead to the simplest way of forming what is called a subgroup.

Names. What's in a name? Every group that you studied has its own name. In fact it has several names. For instance there is the pair $[\mathcal{Z}_{12}, +_{12}]$ which we sometimes call $\mathcal{Z}_{12}$. There is also the name of the operation $+_{12}$. That's not all. Inside the group there is the name for the identity, 0 or zero in this case and the name for the function that assigns to each element its inverse, - or minus .

All that is fine if you are working with a specific group. It is useful to have individual names that help remind you of what various mathematical objects mean. But if there are lots of groups around and you want to work with all of them, or if you just want to talk about a generic group, to say something that will refer to any group whatsoever, then you probably want to use a generic name. The name should not remind you of any particular group, but rather should suggest groups in general.

It is convenient to standardize these generic names. We do this in two ways in this book, the **ISETL** way and the mathematics way. That's not so good. It would be better if we used just one. One reason we can't is that **ISETL**, being a computer system, will not tolerate any ambiguity. Mathematicians are mostly human and can allow an occasional abuse of language when it is convenient. So there has to be a difference. But one of the advantages of **ISETL**, as opposed to other computer languages, is that it does allow us to use standard notation that is pretty close to conventional mathematical notation.

Standard Names. The standard names in **ISETL** are created by the proc name_group. It assigns the following standard notation for a given group.

G - the set

o - the operation (so that a .o b is the "product" of a and b.)

e - the identity element

i - the inverse function (so that $i(g)$ is the inverse of g.)

As we introduce new ideas in what follows, additional standard names will be added to the list.

The role of names. Don't get the idea that these names are more important than the ideas. According to one well-known user of words, there is often difficulty in learning a new word, not because of its sound, but because of the concept to which the word refers: "There is a word available nearly always when the concept has matured". [4] Use the names as a convenience if you like, but learn the concepts first.

What is really important in all this discussion is the relation between the general and specific that these names formalize. Let's illustrate that relation with an analysis of the cancellation law.

2.3.3   The cancellation law—An illustration of the abstract method

The cancellation law is the following statement:

$$\text{If } a \circ b = a \circ c \text{ then } b = c.$$

(this is actually the left cancellation law. Similarly there is a right cancellation law.) In the activities, you investigated examples of the two relations $a \circ b = a \circ c$ and $b = c$. Did you find that whenever the first held, the second was also true? We want to prove this for any three elements in any group. First we'll analyze how it goes in an example, say S_4. We could do this again by checking all possible cases (perhaps using **ISETL**). Or we could prove it more elegantly using the fact that each permutation has an inverse. Recall that one way of stating the relationship between a permutation a and its inverse a^{-1}, is that a^{-1} undoes whatever a does, or, in symbols: for each i, $a^{-1}(a(i)) = i$. Thus we can use the following chain of deductions to prove the cancellation law for S_4:

$a \circ b = a \circ c$	
$a \circ b(i) = a \circ c(i)$ for $i = 1, \ldots, 4$	(definition of equality of permutations)
$a(b(i)) = a(c(i))$ for $i = 1, \ldots, 4$	(definition of the operation $\circ$ in S_4)
$a^{-1}(a(b(i))) = a^{-1}(a(c(i)))$ for $i = 1, \ldots, 4$	(applying the inverse a^{-1} on both sides)
$b(i) = c(i)$ for $i = 1, \ldots, 4$	(property of inverse)
$b = c$	(definition of equality of permutations)

However, there is another, more abstract, way of characterizing the relationship between a and its inverse, that is, $a \circ a^{-1} = e$, the identity permutation. Using this, we can obtain a slightly different proof that is at

---

[4] L. Tolstoy *Pedagogicheskie statli* [Pedagogical Writings], Moscow: Kushnerev, 1903.

once simpler and (as we shall presently see) more general:

$a \circ b = a \circ c$
$a^{-1} \circ (a \circ b) = a^{-1} \circ (a \circ c)$ (operating with the inverse a^{-1}
  on both sides)
$(a^{-1} \circ a) \circ b = (a^{-1} \circ a) \circ c$ (associativity)
$e \circ b = e \circ c$ (abstract property of the inverse)
$b = c$ (property of the identity).

What makes this last proof particularly interesting is the realization that it uses no special properties of permutations. In fact, the same proof could be carried out just as easily using only the group axioms, thereby validating it for any group. In other words, the cancellation law can now be elevated to the rank of a theorem of group theory.

Theorem 2.4 (Groups have the cancellation property.) *Let G be any group, a, b, c any elements in G.*

If $a \circ b = a \circ c$ then $b = c$.
If $b \circ a = c \circ a$ then $b = c$.

Proof. We prove the first assertion (left cancellation law). The second assertion is proved similarly. We start by assuming that $a \circ b = a \circ c$ for some $a, b, c \in G$. We then wish to show that $b = c$. This is established through the following sequence of deductions:

$a \circ b = a \circ c$ (assumed)
$a^{-1} \circ (a \circ b) = a^{-1} \circ (a \circ c)$ (operating with the inverse a^{-1}
  on both sides)
$(a^{-1} \circ a) \circ b = (a^{-1} \circ a) \circ c$ (associativity)
$e \circ b = e \circ c$ (defining property of the inverse)
$b = c$ (defining property of the identity).

□

2.3.4 How many groups are there?

Infinitely many, of course. We saw that $[\mathcal{Z}_n, +_n]$ is a group for any positive integer n. Each of these groups has n elements. Moreover, the group $[\mathcal{Z}, +]$ is infinite. Thus we have a simple but striking result: There are groups of all possible sizes, including infinity.

The next question is, how many of each size? Just one? two? infinitely many? To deal with this question (we will only discuss it for some small numbers), we need some more terminology.

Definition 2.10 (order of a group.) *A group is said to be* finite *or in-finite according to whether it has a finite or infinite number of elements.*

If the group is finite, the number of elements in it is called the order *of the group. If the group is infinite, then its order is said to be infinite.*

We have seen that $\mathcal{Z}_n$ and $\mathcal{Z}$ give us examples of groups of any order (finite or infinite). We now deal with the question "how many of each order?". Let's look at some data. Obviously there is only one group of order 1, the group consisting of the identity all by itself. In the activities you investigated this question for $n = 2, 3$ using the func all_groups. What did you get? Think about the following two-by-two "multiplication tables".

o	e	1
e	e	1
1	1	e

o	e	-1
e	e	-1
-1	-1	e

o	e	a
e	e	a
a	a	e

$\mathcal{Z}_2$ G Arbitrary

Right away there is a problem of interpretation. Clearly in the above question we mean "how many *different* groups of each order?". But is it so obvious what is meant by two groups being different or the same? For example, in the past you have seen at least two groups of order 2, $\mathcal{Z}_2$ with addition mod 2 and the group $\{1, -1\}$ with ordinary multiplication. What would happen if you ran these two groups through name_group and displayed the operations as in the tables above? (We have added a third one to represent an arbitrary group of order 2.) Well, first there is the question of names. The proc name_group would call the two operations o and the identity elements of the two groups e. The remaining elements would be 1 for $\mathcal{Z}_2$ and -1 for G.

Now, except for the different name given to the second element, would you say that these three groups are different or the same? Well, at least in the context of finding all groups of a certain order, we'll consider them the same: each one is just a renaming of the others. We say that

the groups $[\{1, -1\}, *]$ and $[\mathcal{Z}_2, +_2]$ are isomorphic.[5]

(And so is the third one isomorphic to each of them.) It means that we can move from one to the other simply by changing the names. In general if you try to fill in a two-by-two table to satisfy the group axioms, then you see that there is really only one way to do it. In this sense, there is only one group of order 2. The only issue is the name that is given to the second element. (Incidentally, this interpretation was tacitly assumed also when we said there was only one group of order 1.) This is expressed mathematically as a proposition.

---

[5]A more formal definition and a more complete discussion of isomorphism will be given in Section 4.2.6.

Proposition 2.17 *All groups of order 2 are isomorphic to the group* Z_2.

Proof. Exercise 5.

What do you think about groups of order 3?

Classifying groups of order 4

We know there is at least one, $[Z_4, +_4]$. We'll refer to it as Z_4 for short. To find others, let's look at the multiplication table and see how to fill it in while keeping the group axioms in mind. Associativity is a killer, so let's ignore that for a while. Let's name the elements e, a, b, c where e is the identity and the others are different from each other and from e. As with order 2, the property of the identity permits us to fill in the first row and column immediately.

case 1

	e	a	b	c
e	e	a	b	c
a	a		e	
b	b			
c	c			

case 2

	e	a	b	c
e	e	a	b	c
a	a			c
b	b			
c	c			

	e	a	b	c
e	e	a	b	c
a	a			
b	b			
c	c			

Let's work on the second row. Ignoring a^2 for a moment, what possibilities are there for the value of ab? The closure axiom tells us that it must be one of e, a, b or c. It can't be that $ab = a$ for this would imply $b = e$ (apply the cancellation law to $ab = ae$), which is false. Similarly, we can't have $ab = b$. The only possibilities are $ab = e$ or $ab = c$. Lets make two charts and pursue these separately.

The same kind of argument we just made using closure and the cancellation law leads to the following general fact.

Remark. In a group chart, each element appears exactly once on any row or column.

This lets us fill in the table in case 1 pretty quickly. Can you fill in the details, making sure that each of your decisions is compatible with the group axioms and is, in fact, the only one possible? You should come up with the following table:

case 1

	e	a	b	c
e	e	a	b	c
a	a	c	e	b
b	b	e	c	a
c	c	b	a	e

+4	0	1	2	3
0	0	1	2	3
1	1	2	3	0
2	2	3	0	1
3	3	0	1	2

We are not yet sure that we have a group, but notice that we have placed the table for the group $\mathcal{Z}_4$ next to it. Are they the same? Does the renaming trick work? Well, perhaps if we do it carefully. Of course e is the identity and that corresponds to 0. Naturally we try to let a correspond to 1, but what about b and c? Already we can see that they cannot correspond to 2 and 3 in this order. This is because the elements in the third row and the second column of the present tables are e and 3 and hence cannot correspond (e is already matched with 0). We can try to reorder b and c, that is, correspond c to 2 and b to 3. Here is how the tables look now:

case 1

	e	a	c	b
e	e	a	c	b
a	a	c	b	e
c	c	b	e	a
b	b	e	a	c

+4	0	1	2	3
0	0	1	2	3
1	1	2	3	0
2	2	3	0	1
3	3	0	1	2

Now they are the same. Once you translate the names of elements, the results of the operations (that is, the tables) translate consistently. The group we have constructed is just a renaming of the group $\mathcal{Z}_4$, that is, it is isomorphic to $\mathcal{Z}_4$. Note however that, as we have seen, renaming may involve reordering. Incidentally, since the two tables are the same except for names, the fact that $\mathcal{Z}_4$ is a group implies that what we have constructed is also a group.

So, maybe all groups of order 4 are isomorphic to $\mathcal{Z}_4$? We still have case 2 to consider, that is, filling in the table assuming $ab = c$. To finish the row for a we need to determine the value of ac. As before, ac cannot be a or c. It could be e, but then by going back and renaming a as c and c as a, we would get a group isomorphic to that of case 1 and therefore to $\mathcal{Z}_4$. Thus in our search for new groups we may assume $ac = b$. Then we get (again leaving the details for you) the following multiplication table of the Klein group:

	e	a	b	c
e	e	a	b	c
a	a	e	c	b
b	b	c	e	a
c	c	b	a	e

You can use is_associative to check associativity and the table guarantees that the other axioms are satisfied (do you see this?). So it is a group[6]. The table also guarantees that this is a commutative group. (Can you see how this is expressed in the form of the table?) But is it isomorphic to Z_4? The answer is no. Why not? Think about it.

By way of a hint, we will look at two other groups of the same order, S_3 and Z_6, and show that they are not isomorphic. The idea is to show that they differ in some property that should be the same if they had the same table. (Such properties are called invariants.) Indeed S_3 and Z_6 cannot have the same table except for renaming, since Z_6 is commutative while S_3 is not. Now can you use this kind of reasoning to show that our last group of order 4 is not isomorphic to Z_4? This new group of order 4 is called the *Klein 4-group* (after the German mathematician Felix Klein) and is denoted K. The above discussion can then be summarized in the following result.

Theorem 2.5 *There are exactly two groups of order 4. More precisely, any group of order 4 is isomorphic to either Z_4 or to K.*

Proof. Exercise 7.

2.3.5   Looking ahead—subgroups

In the activities you investigated what happens if you take a subset of a group and let it inherit the operation from the original group. Sometimes you get a group and sometimes you don't. There is no general rule, but we can pretend we understand everything by making a definition.

Definition 2.11 *Let $[G, \circ]$ be a group and let H be a subset of G. If $[H, \circ]$ is again a group, then we say that H is a subgroup of G.[7]*

---

[6]For those of you who are interested in psychology, this is none other than the INRC group of J. Piaget. We have e is I(dentity), a is N(egation), b is R(eciprocity) and c is C(orrelation).

[7]Strictly speaking, we should write $[H, \circ_H]$ to indicate the restriction of the operation $\circ$ to H.

Example: Recall the group D_4 of all the symmetries (or repositionings) of the square, which we discussed on p. 49. Altogether it contains 8 symmetry transformations. Let us look at the subset K' of D_4, consisting of the identity transformation, the horizontal and the vertical flips, and the 180-degrees rotation; in symbols

$$K' = \{R_0, R_2, H, V\}.$$

Use the chart on p. 51 to build a multiplication table for K', and convince yourself that it is a subgroup of D_4. Furthermore, put this multiplication table side-by-side with that of K (p. 79) and convince yourself that K' and K are isomorphic. We might conclude that the Klein 4-group K, which was originally introduced here in a very abstract setting, has found now a concrete, geometric realization.

Here is a general method for constructing subgroups.

Proposition 2.18 *Let g be an element of a group G and let H be the set of all powers of g, positive, negative, or 0. Then H is a subgroup of G.*

Proof. Exercise 9.

The group H of the previous proposition is called the *subgroup generated by g* and is denoted $\langle g \rangle$.

2.3.6 *Summary of examples and non-examples of groups*

Where the choice of operation is unambiguous, we will write only the set.

System	Group?	Axiom violated
$[\mathcal{Z}, +]$	Yes	
$[\mathcal{Z}, *]$	No	Inverses
$[\mathcal{Z} - 0, *]$	No	Inverses
$[\mathcal{Q}, +]$	Yes	
$[\mathcal{Q}, *]$	No	Inverses
$[\mathcal{Q} - 0, *]$	Yes	
$[\mathcal{Z}_n, +_n]$	Yes	
$[\mathcal{Z}_n, *_n]$	No	Inverses
$[\mathcal{Z}_n - \{0\}, *_n]$ n prime	Yes	
$[\mathcal{Z}_n - \{0\}, *_n]$ n not prime	No	Closure
$[\mathcal{Z}_n, +]$	No	Closure
S_n	Yes	
D_4	Yes	
$[\mathcal{R}, +]$	Yes	

System	Group?	Axiom violated
$[\mathcal{C}, +]$	Yes	
K	Yes	
2×2 matrices over $[\mathcal{R}, +]$	No	Inverses
2×2 non-singular matrices $[\mathcal{R}, +]$	Yes	

2.3.7 EXERCISES

1. Suppose that G is a non-empty set on which a closed, associative operation ("product") is defined. Prove or give a counter example to the statement that if every equation $ax = b$ with $a, b \in G$ has a solution in G, then G is a group.

2. Prove that if G is a *finite* set on which a closed, associative operation ("product") is defined which satisfies both left and right cancellation laws, then it is a group. Give examples to show that if only one cancellation law holds or if G is infinite, then G may fail to be a group.

3. Suppose that G is a set on which a closed, associative operation ("product") is defined and that G has an identity element relative to this operation. Prove that the subset of all invertible elements of G is a group.

4. Prove that in a group chart, each element appears exactly once on any row and exactly once on any column.

5. Show in detail that there is only one way to fill in the following tables so that we get a group (assume e, a, b are all different from each other):

	e	a
e		
a		

	e	a	b
e			
a			
b			

Deduce that all groups of orders 2 and 3 are isomorphic to $\mathcal{Z}_2$ and $\mathcal{Z}_3$, respectively.

6. Complete the details of the calculations leading up to the group-table on p. 78. Prove that the group of order 4 defined by that table is *not* isomorphic to $\mathcal{Z}_4$.

7. Complete the proof of Theorem 2.5, that is, any group of order 4 is isomorphic to either $\mathcal{Z}_4$ or to the Klein 4-group K.

8. How many different (i.e. non-isomorphic) kinds of groups of order 5 are there?

9. Prove Proposition 2.18, that is, let g be an element of a group G and let H be the set of all powers of g, that is, $H = \{g^n : n \in \mathcal{Z}\}$. Then H is a subgroup of G.

10. If G is a finite group of order n, prove that we can get the subgroup H generated by g (see previous exercise) more economically as $H = \{g, g^2, \ldots, g^n\}$.

11. Find a subgroup of order 4 in S_4. Can you find a subgroup of order 4 in S_3?

12. Formulate and prove the "generalized associative law" for groups. Informally this law says that no matter how you bracket a "generalized product" $a_1 a_2 \ldots a_n$ in a group, you always get the same result. For example, $(a_1(a_2 a_3))a_4 = (a_1 a_2)(a_3 a_4)$.

3

Subgroups

3.1 Definitions and examples

3.1.1 ACTIVITIES

1. Edit your **proc** name_group to obtain a new **proc** group_table which will define the same names G,o,e,i as name_group, but where o and i will be maps rather than funcs. For example, o will be the set of all pairs of the form [[x,y], x .o y] . Note that expressions like a .o b and i(g) should evaluate to the same group elements as with name_group. Evaluating group_table on a particular group may take much longer than name_group, but once completed, investigations with that group will be much faster.

2. Describe the connection between the revision of name_group in the previous activity and the multiplication tables you worked with in Chapter 2, Section 3.

3. At the end of Chapter 2 the concept of subgroup was described. Use your **func** is_group to check if H is a subgroup of G in the following.

   (a) G = Z20, H is the set of even integers in G.

   (b) G = S4, H is the set of permutations whose support (see p. 58 and p. 68) does not contain 4.

   (c) G = S3, H = A3 = {(1), (12)(13), (12)(23), (13)(23)}.

   In each case for which the answer is no, explain why.

4. Write a func is_subgroup that assumes name_group has been run (so you can use G,o,e,i), accepts a set H and checks if H is a subgroup of G. Use the most efficient method for checking that you have found. Check if H is a subgroup of G in the following. (In each of these, the operation is understood to be the usual one that makes G a group.)

   (a) G = Z11 - {0}, H = {0, 1, 2, 3}

   (b) G = Z12, H = {0, 1, 2, 3}

   (c) G = Z12, H = {0, 3, 6, 9}

   (d) G = Z20, H = {0, 3, 6, 9}
       (See Section 2.2.1, p. 57 and Section 2.2.3, p. 65.)

   (e) G = S4, H is the set of all powers of the cycle (1 2 3).

   (f) G = S4, H is the set {(1 2), (1 3), (1 4)}.

   (g) G = S4, H is the set of all powers of the permutation (1 2)(3 4).

   (h) G = S4, H = {(1), (1 2), (3 4), (1 2)(3 4)}.

   (i) G = S4, H = {(1), (1 2)(3 4), (1 3)(2 4), (1 4)(2 3)}.

   (j) G = S4, H = {p : p in S4 | p(4) = 4} (compare with Activity 3(b)).

   (k) G = S4, H = {p : p in S4 | p(4) = 3}.

   (l) G = S4, H is the set of all permutations which can be written as a product of an even number of transpositions. (A transposition is a cycle of length 2.)

   (m) G = S4, H is the set of all permutations which can be written as a product of an odd number of transpositions.

5. One particular method of calculation that can be very efficient is to use **VISETL**, that is, your mind. This is the only way when the sets are infinite. Check if H is a subgroup of G in the following.

   (a) The even integers in Z.

   (b) The odd integers in Z.

   (c) The set 3Z of all multiples of 3 in Z.

   (d) The set 3Z+1 of all integers that leave a remainder of 1 on division by 3.

   (e) The integers in Q.

   (f) The integers in Q - {0}.

6. For each positive integer j from 1 to 12, try to find a subgroup of Z_{12} which is of order j.

7. Is there a subgroup of S_4 which is isomorphic to the Klein group, K? (See p. 79.) Recall that by "isomorphic" we mean that it has the same operation table except for reordering and renaming.

8. Is D_4 isomorphic to some subgroup of S_4? Is S_3?

9. In this activity, you may want to use the **ISETL** func **pow** which accepts a set S and returns its power set, that is, the set of all subsets of S.

   Write a func that will accept a group and produce the set of all of its subgroups. Run your func on the following.

   (a) Z12,

   (b) Z11,

   (c) S3.

10. Based on all of the examples and non-examples you have seen, do you think there is a relationship between the order of a group and the orders of its subgroups?

11. Take any example you like of a group G, a subgroup H and an element $g \in G$. Let
$$gHg^{-1} = \{ghg^{-1} : h \in H\}$$
Is gHg^{-1} again a subgroup of G? The answer might depend on the particular G, H, g. Look for an example in which the answer is yes and one in which it is no. The set gHg^{-1} is called the *conjugate* of H by g.

12. Repeat the previous activity with gHg^{-1} replaced by
$$Hg = \{hg : h \in H\}$$
Sets of the form Hg are called *right cosets* of H.

13. If g, h are elements of a group G we also consider the element
$$ghg^{-1}$$
called the conjugate of h by g. The following can be done with paper and pencil.

   (a) Express the conjugate of a product ab by an element c in terms of the conjugates of a and b by c.

   (b) What is the conjugate of the identity of a group?

   (c) What is the conjugate of the inverse of an element?

14. Calculate the following conjugates in S_4:

(a) $(1\,2\,3)$ by $(3\,4)$.

(b) $(1\,2\,3)$ by $(2\,3)$.

(c) $(1\,2)$ by $(3\,4)$.

(d) $(1\,2)(3\,4)$ by $(2\,3)$.

(e) $(1\,3)(2\,4)$ by the permutation,

$$\begin{pmatrix} 1\,2\,3\,4 \\ 1\,3\,2\,4 \end{pmatrix}.$$

Based on these examples, can you state what a conjugation does to a cycle? To a product of disjoint cycles?

15. Calculate the conjugates of $(1\,2\,3\,4)$ by $(3\,4)(2\,3)$ and

$$\begin{pmatrix} 1\,2\,3\,4 \\ 1\,3\,2\,4 \end{pmatrix}.$$

16. If p, q are permutations and if p sends a to b, what can you say about the action of qpq^{-1} on $q(a)$? Write out a general statement describing this situation.

3.1.2 Subsets of a group

The main purpose of this section is for you to become familiar with examples of subsets of groups that may or may not be subgroups. There are two main reasons why we study subgroups. First, they provide a rich source of examples of groups. And second, we want to see how a group, like any edifice, is a complex whole that is made up of simpler parts. For this study of the "structure" of groups, subgroups are important building blocks.

In the activities you saw two ways in which one subset of a group can be transformed into another: conjugation and cosets.

Conjugation of subgroups provides examples of more subgroups and we will discuss them here. Cosets actually don't (did you discover this in the activities?) but will contribute in other important ways to an investigation into the structure of groups. They will be dealt with in the last section of this chapter. The information you obtained about orders of groups and their subgroups will provide, in the last section of this chapter, a surprising insight into what groups and subgroups are possible.

Definition of a subgroup

A major source of interesting and important groups are the subsets of a given group. Given a group $[G, o]$, you can always take a subset S of the group G and form the system $[S, o]$. Here o is the group operation which S inherits from G, but considered to operate only on pairs of elements from

S. As you have seen in the activities it can go both ways. In some examples, $[S, \circ]$ is a group and in some it is not. We repeat the definition.

Definition 3.1 (subgroup.) *Let $G = [G, \circ]$ be a group and let H be a subset of G. If $[H, \circ]$ is a group then we say that H is a* subgroup *of G.*

Strictly speaking, $\circ$ is not *exactly* the same as the group operation on G. It is the restriction of this operation to H. There is no danger of confusion, however, and we will generally use the same symbol to denote both.

Remark: As you presumably discovered from the activities, when testing for a subgroup you don't have to really check the group properties in the same way that you determined whether a given set with a binary operation is a group. Of course you have to check that the given set is a subset of the set in the original group. Then you must check closure. Associativity, however, is inherited by all subsets, hence can be taken for granted. Also you don't have to look for the identity and for inverses. It turns out that the only possible identity is the identity of the original group. So the only thing you have to check is that the identity of the original group is in the subset. Similarly for inverses. Thus we have the following result.

Proposition 3.1 (tests for subgroups.) *Let G be a group and H a subset of G. Then H is a subgroup of G if and only if the following three conditions are satisfied:*

1. *H is closed under the operation induced by the group G.*

2. *The identity of G is in H.*

3. *The inverse (in G) of every element of H is again in H.*

Proof. The proof that any H which satisfies these conditions is a group is immediate because the only thing that is missing is associativity which is inherited. The converse also seems immediate but there is one sticky point. Could it be that the identity e for G is not in H, but still H is a subgroup? Perhaps there could be another element of H that serves as the identity? This can't happen. We leave the explanation for you to give in the exercises.

$\square$

As a practical matter, the test in Proposition 3.1 represents a real reduction in the time required to check for subgroups on the computer, because it eliminates the need to check associativity which takes longer than any other of the group axioms. Also instead of computations to decide which element is the identity or the inverse of a given element, this proposition tells us that there is only one candidate, the identity in G or the inverse in G of the element.

From the activities, you should have a long list of examples of subgroups. You should also know how to find many more. You should have no trouble seeing that the group itself and the singleton subset consisting of the identity alone satisfy the definition of subgroup. These have a special name.

Definition 3.2 (proper and trivial subgroups.) *If G is a group then the subgroup $\{e\}$ is called the* trivial *subgroup of G. All other subgroups other than G itself are called* proper *subgroups.*

We single out two subgroup examples of particular interest - those appearing in Activities 4(h) and 4(i). The proof of the following statements is left as an exercise.

Example. The sets

$$K_1 = \{(1), (12), (34), (12)(34)\}$$

and

$$K_2 = \{(1), (12)(34), (13)(24), (14)(23)\}$$

are subgroups of S_4, both isomorphic to the Klein 4-group (hence also to each other). Thus we have two new "realizations" of Klein's group, this time as permutation groups. However, if you think of the digits $1, 2, 3, 4$ as marking the corners of a square (say clockwise from the top-left corner), then you can see that K_2 is essentially the same as the subgroup K' of symmetries of the square that we discussed in the example just preceding Proposition 2.18, p. 80. Can you find another subgroup of D_4 (perhaps after re-numbering the corners) that is "esentially the same" as K_1 and K_2?

3.1.3   Examples of subgroups

Embedding one group in another

There is a subtle point about subgroups and subsets that we would like to emphasize. In Activity 4(b), you considered the subset $H = \{0, 1, 2, 3\}$ as a subset of Z_{12}. Now H is the same set as Z_4 which is a group but, hopefully, you decided that H is *not* a subgroup of Z_{12}. The point is that there are two operations we might consider on this set H. One is $+_4$ which comes from the group Z_4 and the other is $+_{12}$ (restricted to H) which it inherits from Z_{12}. So we have a case of two operations defined on the same set. It is a group with respect to the first, $+_4$, but not with respect to the second, $+_{12}$. Another way of saying this is that Z_4 is a group but it is not a *sub*group of Z_{12}.

Now let's make things a little more complicated. Still thinking about Z_{12}, what about the subset $H = \{0, 3, 6, 9\}$? This was one from the activities that turns out to be a subgroup. So it is a group with 4 elements. Recall from Chapter 2, Theorem 2.5, p. 79, that any group with four elements is isomorphic to, i.e., is a renaming of either $\mathcal{Z}_4$ or K. Which is it? Make a quick check in your head. Right, it's $\mathcal{Z}_4$. So $\mathcal{Z}_4$ is isomorphic to a subgroup of $\mathcal{Z}_{12}$.

Do the last two paragraphs contradict each other? That is, is there any contradiction in saying on the one hand that $\mathcal{Z}_4$ is not a *subgroup* of $\mathcal{Z}_{12}$ and on the other hand that $\mathcal{Z}_4$ is isomorphic to a subgroup of $\mathcal{Z}_{12}$? Look carefully at the wording and satisfy yourself that there is no contradiction.

It is sometimes very useful to know that a group can be *embedded* in another group. That is, it can be identified as being isomorphic to a subgroup of another group. It can even happen (as we will see below) that a group can be embedded in itself as a proper subgroup. Here are a few classes of examples. As usual, we leave all the fun of proving them for you in the exercises.

Proposition 3.2 (embedding S_{n-1} in S_n.) *For every integer $n > 1$ the group S_{n-1} is isomorphic to a subgroup of S_n.*

Proof. Exercise 7.

For example, S_3 can be embedded in S_4.

Proposition 3.3 (embedding $\mathcal{Z}_k$ in $\mathcal{Z}_n$.) *If n is a positive integer and k is a divisor of n, then $\mathcal{Z}_k$ is isomorphic to a subgroup of $\mathcal{Z}_n$.*

Proof. Exercise 8.

For example, $\mathcal{Z}_4$ can be embedded in $\mathcal{Z}_{12}$.

Proposition 3.4 (embedding $\mathcal{Z}$ in $\mathcal{Z}$.) *For every positive integer n the set $n\mathcal{Z}$ of multiples of n is a subroup of $\mathcal{Z}$. It is isomorphic to $\mathcal{Z}$ itself.*

Proof. Exercise 9.

For example, $\mathcal{Z}$ is isomorphic to its own subgroup $3\mathcal{Z}$.

Conjugates

Conjugation is an operation which can be performed on both a subgroup and a single element. You have worked with both in Activities 11-15, pp. 85-86. We will discuss the two versions in parallel.

Definition 3.3 (conjugates.) *Let G be a group and $g, x \in G$. The* conjugate *of x by g is the element gxg^{-1}.*

If H is a subgroup of G the conjugate of H by g, *written gHg^{-1}, is the set of all conjugates of elements of H by g. That is,*

$$gHg^{-1} = \{gxg^{-1} : x \in H\}.$$

The following proposition collects some elementary facts about conjugates, some of which you have already noticed in the activities.

Proposition 3.5 (properties of conjugates.)

1. *A conjugate of a subgroup is a subgroup.*

2. *A conjugate of a subgroup has the same order as the original subgroup.*

3. *A conjugate of the product of two elements of a group is the product of their conjugates.*

4. *A conjugate of the identity element is the identity element.*

5. *The inverse of a conjugate of an element is the conjugate of its inverse.*

Proof. We leave as an exercise both a precise formulation of this proposition and the proof.

Remark. In a sense, conjugating by g is just renaming. (We rename x as gxg^{-1}.) It can therefore be expected that H and gHg^{-1} should be isomorphic. (We shall discuss this issue more fully later.) It is no wonder, then, that so many properties are "preserved".

We switch now from considering conjugation by a fixed element to generic conjugates.

Definition 3.4 *An element $a \in G$ is said to be* conjugate *to $b \in G$ if there is an element $c \in G$ such that $b = cac^{-1}$. Similarly, one subgroup is said to be* conjugate *to another if there is an element of the group such that the second subgroup is a conjugate of the first by the element.*

Things which are conjugates have many properties in common. In fact, for some purposes, they behave as equals. The next proposition shows that conjugation indeed has some of the basic properties of equality.

Proposition 3.6 *Conjugation of elements and of subgroups both form equivalence relations. More specifically (for elements) we have the following.*

1. *Any element of a group is conjugate to itself.*

2. *If a is conjugate to b then b is conjugate to a.*

3. *If a is conjugate to b and b is conjugate to c then a is conjugate to c.*

Proof. Exercise 11.

Cycle decomposition and conjugates in S_n

We now turn from the general properties of conjugation to a more intimate study of what conjugation looks like in some specific cases. You should see right away that conjugation is not very interesting in commutative groups (why?) and so we turn to our main class of non-abelian groups - the symmetric groups S_n.

In Activities 15 and 16 you computed conjugates pap^{-1} of elements $a \in S_4$ by elements $p \in S_4$, where a was written as a product of disjoint cycles. Did you notice that you can read this answer off immediately by just replacing each number i in a by $p(i)$? This always happens for permutations written in this form and, in fact, any permutation can be so written. That is, there are two general results here. First, that any permutation can be written as a product of disjoint cycles, and second, that conjugating a product of disjoint cycles by p can be done by just applying p to the numbers in the cycles.

Before considering the first result we recall a closely related activity you did in Section 2.2.1 Activities 8 and 9, p. 58 - decomposition into *orbits*. In Activity 8, you wrote the orbits of

$$p = \begin{pmatrix} 1\,2\,3\,4\,5\,6\,7\,8 \\ 5\,6\,8\,2\,1\,4\,7\,3 \end{pmatrix}$$

as follows:

$$1 \to 5 \to 1$$
$$2 \to 6 \to 4 \to 2$$
$$3 \to 8 \to 3$$
$$7 \to 7$$

This decomposition of p into its orbits has a parallel decomposition in terms of cycles, namely, as a *product* of the corresponding cycles: $p = (1\ 5)(2\ 6\ 4)(3\ 8)$. (check that the two sides are indeed equal).

Theorem 3.1 (cycle decomposition.) *Every permutation (other than the identity) can be written as a product of disjoint cycles.*

Discussion of the proof. We have seen above how the decomposition is actually carried out: you write out the orbits, then the corresponding cycles. The complete proof essentially consists in formulating the decomposition process in general terms and in showing that this can *always* be done. In particular you'll have to use the 1-1 property of permutations to show that (a) the first repetition in an orbit (or cycle) is always the opening number, and (b) that different orbits (or cycles) arising in this way are disjoint. Finally you'll have to show that the original permutation is indeed equal to the *product* of the cycles arising in this way. We leave these details for the exercises.

Remark. The decomposition of Theorem 3.1 is essentially unique. The precise formulation as well as the proof of this remark are left for the exercises.

Remark. Had we not decided that the identity permutation should not be considered a cycle, this would give rise to a source of non-uniqueness, for the identity can be written as $(1) = (2) = \dots$.

Now we come to the second result.

Proposition 3.7 (conjugate of a permutation.) *If a is a permutation, represented as a product of disjoint cycles, then the conjugate pap^{-1} of a by p is obtained by replacing each number i appearing in the cycles by $p(i)$.*

Proof. First consider the case in which $a = (u\ v\ w\ \dots)$ is a cycle. We wish to show that $pap^{-1} = (p(u)\ p(v)\ p(w)\dots)$. It is enough to check that these two permutations agree on the numbers $p(u), p(v), p(w), \dots$ (why?). Indeed it is clear that the cycle on the right hand side sends $p(u)$ to $p(v)$, while for pap^{-1} we calculate,

$$pap^{-1}(p(u)) = pa(u) = p(v)$$

and similarly for the others. Thus the two permutations are the same. In general, if a is the product of cycles, then we apply this result to each cycle and invoke the proposition on the conjugate of a product.

$$\square$$

Thus, in the sense of this proposition, we can say that *conjugate permutations have the same cycle structure.*

The cycle decomposition theorem has many other uses. We will use it to determine two important properties of a permutation: order and parity (to be introduced later in this chapter).

3.1.4   EXERCISES

1. Complete the proof of Proposition 3.1; that is, show that if H is a subgroup of G then the three conditions listed in the proposition are indeed satisfied.

2. Recall that Klein's 4-group (p. 79) may be defined as the group with 4 elements $K = \{e, a, b, c\}$ where the multiplication is defined by the following relations: the square of each element is e, the identity element, and the product of any two non-identity elements (in either order) is equal to the third such element. Show that the subsets K_1 and K_2 of S_4 defined on p. 88 are subgroups and are isomorphic to K. Find all subgroups of S_4 which are isomorphic to K.

3. Let G be the group of all real matrices $\begin{pmatrix} a & b \\ c & d \end{pmatrix}$ (see p. 52) satisfying $ad - bc \neq 0$. Let S be the subset of all elements of G satisfying $ad - bc = 1$ and T those of the form $\begin{pmatrix} \cos x & \sin x \\ -\sin x & \cos x \end{pmatrix}$. Show that both S and T are subgroups of G and that T is a subgroup of S.

4. Let M_n be the set of all non-singular (i.e., invertible) $n \times n$ matrices with real entries. Show that M_n is a group.

5. In the context of the previous exercise, check each of the following subsets of M_n for being a subgroup.

   (a) The subset of all diagonal matrices (all entries not on the main diagonal are zero) in M_n.

   (b) The subset of all lower triangular matrices (i.e., every element above the main diagonal is zero) in M_n.

   (c) The subset of all lower triangular matrices in M_n in which all of the entries on the main diagonal are 1.

   (d) The subset of all lower triangular matrices in M_n in which all of the entries on the main diagonal are either 1 or -1.

   (e) The subset of all lower triangular matrices in M_n in which all of the entries on the main diagonal are non-zero integers.

   (f) The subset of all lower triangular matrices in M_n in which all of the entries on the main diagonal are non-zero rational.

   (g) The subset of all lower triangular matrices in M_n in which all of the entries on the main diagonal are elements of some fixed subgroup of $[\mathcal{R} - \{o\}, *]$.

6. How much of the previous two exercises can you generalize by replacing the non-zero reals with an arbitrary group?

7. Prove Propositions 3.2.

8. Prove Propositions 3.3.

9. Prove Propositions 3.4.

10. Prove Propositions 3.5.

11. Prove Propositions 3.6.

12. Is there a subgroup of $\mathcal{Z}$ isomorphic to $\mathcal{Z}_5$?

13. (J. Gallian) Given that $[\{1, 9, 16, 22, 53, 74, 79, 81, x\}, *_{91}]$ is a group, determine the number x.

14. (J. Gallian) H is a subset of $\mathcal{Z}_{28}$ which forms a group of order 12 under multiplication mod 28. If it is known that $5, 15 \in H$, determine all the elements of H.

15. Let n be a fixed positive integer, RU the set of all complex numbers satisfying $z^n = 1$ (these are called the *complex n^{th} roots of unity*). Show that RU is a group under ordinary multiplication of complex numbers.

16. Prove that the set of all real numbers of the form $a + b\sqrt{2}$, where a, b are rational numbers which are not both 0, forms a subgroup of the multiplicative group of non-zero real numbers.

17. Complete the proof that every permutation can be written as a product of disjoint cycles (Theorem 3.1).

18. Formulate and prove a "uniqueness" theorem for the decomposition of a permutation into a product of disjoint cycles. (You'll have to take into account two sources of non-uniqueness, as demonstrated in the following two equalities: $(1\ 2\ 3) = (2\ 3\ 1)$ and $(1\ 2)(3\ 4) = (3\ 4)(1\ 2)$.)

19. Formulate and prove the assertion in the proof of Proposition 3.7 to the effect that it is enough to check agreement on the given numbers.

20. For any group G define $Z(G) = \{z \in G : za = az \text{ for all } a \in G\}$ ($Z(G)$ is called the *center* of G). Prove that $Z(G)$ is an Abelian subgroup of G.

21. Determine the centers of the groups $\mathcal{Z}_n$, S_n and the matrix groups of Exercise 3.

22. Suppose S and T are subgroups of a group G.

   (a) When is $S \cap T$ also a subgroup of G? Prove your statement.

   (b) When is $S \cup T$ also a subgroup of G? Prove your statement.

23. Show that if H is a non-empty finite subset of a group G and H is closed under the group operation, then H is a subgroup of G.

24. Complete the details of the proof of Theorem 3.1, p. 91.

3.2   Cyclic groups and their subgroups

3.2.1   *ACTIVITIES*

1. If $[G, o]$ is a group, a an element of the group and n a positive integer, explain what is the meaning of the folowing **ISETL** expression:

$$\text{\% .o [a: i in [1..n]];}$$

(You might want to try out some examples.)

The meaning of this expression depends on one of the group axioms being satisfied. Do you see which?

2. Write an **ISETL** func pG which, assuming that either name_group or group_table has been run, accepts an element g of G and an integer n (positive, negative or 0) and returns the nth power of g. (see p. 53.) Apply your func in the following situations.

   (a) G = Z12, g = 3, n = 1, 2, ...,13.

   (b) G = S4, g = (1 2), n = 1, 2, 3, 4.

   (c) G = S4, g = (1 2 3), n = 1, 2, ...,6

   (d) G = S4, g = (1 2 3 4), n = 3, 4, 5, 6, 10.

   (e) G = S4, g = (1 2)(3 4), n = 1, 2, 3, 4.

3. Write an **ISETL** func gen which, assuming that either name_group or group_table has been run, accepts an element g of G and returns the set of all powers of g.

   Hint. The main step of your func should use pG in a set former, but you need to have some upper bound for the power that you will take. There is a mathematical point here that might help you.

   Apply your func in the following situations.

   (a) G = Z20, g = 9.

   (b) G = Z20, g = 8.

   (c) G = Z20, g = 5.

   (d) G = S4, g = (1 2).

   (e) G = S4, g = (1 2 3).

   (f) G = S4, g = (1 2)(3 4).

4. Use your func is_subgroup to determine for which g in the previous activity is gen(g) a subgroup of G. Can you state and prove a theorem that explains your observations?

5. Determine if there are elements g of Z12 with the property that gen(g) is all of Z12. If so, find them all.

6. Determine if there are elements g of S3 such that gen(g) is all of S3. If so, find them all.

7. Apply gen to each element of Z12. Write down all of the different sets that you get.

    (a) What do you observe about the collection of sets?

    (b) What do you observe about the number of elements in each set in relation to the order of Z12?

    (c) Which of the sets that you obtain are subgroups?

    (d) Identify the subgroups that you obtain as familiar groups. That is, for each subgroup, try to find one of the groups you have been working with which is isomorphic to it.

8. Do the same as the Activity 7 for Z11.

9. Do the same as Activity 7 for S4.

10. Summarize any conclusions that you can draw from Activities 7, 8, 9.

11. Write a **func order** that, assuming that **name_group** has been run, will accept an element **g** and return the number of elements in the set of all of its powers. Apply your **func** to each element of Z12 and of S3. Do you notice anything about the numbers that come out?

12. Apply your **func** of the previous activity to a cycle in S4 and to a product of two disjoint cycles, such as (1 2)(3 4 5) in S5 and (1 2 3 4)(5 6 7 8 9 10) in S10. What is the most general conjecture you can make for the result of applying your **func** to a permutation? Of course, in looking for a conjecture, you should use your **func** to collect data. Then guess some rule and use your **func** to verify it or suggest a better rule.

13. Can you write the cycle (1 2 3 4) as the product of transpositions (i.e. cycles of length 2)? Can you do this in more than one way?

14. What are all the elements of S_4 that you can get by conjugating (1 2) by various elements of S_4?

3.2.2   The subgroup generated by a single element

What happens if you start taking successive powers

$$g^0(= e), \; g^1(= g), \; g^2, \; g^3, \; g^4, \ldots$$

of an element g in a group G ? Does it stop in the sense that you don't get any more new elements? If so, where? Well, it might not stop at all. Suppose you did it with the number 3 in the group $\mathcal{Z}$? (Recall that since $\mathcal{Z}$ is an additive group, the "powers" of 3 are actually the multiples.) This group has an inexhaustible supply of multiples of 3. So it never stops. In other cases it will stop. For example if G is finite, sooner or later, there must be a repetition. It is interesting that once you get a single repetition, you never get anything new. The reason is that the first repetition must be $g^k = e$. This shouldn't be news once you have done the activities for this section, and it will be proved later in this section. The following definition formalizes the possibilities.

Definition 3.5 (order of an element.) *Let g be an element of a group G. The smallest positive integer k such that $g^k = e$ is called the* order *of g. If there is no such positive integer, then we say that g has* infinite order.

What do you think the order of an element of a group has to do with the order of a group, discussed in Chapter 2?

Look back at Activities 2 and 3, where you computed many powers of various elements. From these activities you should have no difficulty in reading off the orders of these elements. For example, what is the order of 3 in $\mathcal{Z}_{12}$? of 4 in $\mathcal{Z}_{12}$? of 5 in $\mathcal{Z}_{20}$? of (1 2), (1 2 3), (1 2 3 4) in S_4? The next proposition generalizes these examples. Can you guess before reading further what it says?

Proposition 3.8 *(a) Let $a \in \mathcal{Z}_n$ and suppose that a divides n. Then the order of a is n/a. (b) The order of a cycle in S_n is the same as its length.*

Proof.

(a) Exercise 1.

(b) Consider a cycle $p = (a_1\ a_2\ a_3\ \ldots\ a_k)$ of length k. Note that the numbers $a_1, a_2, \ldots, a_k, a_1$ are just what you get if you apply the sequence of powers $e, p, p^2, \ldots, p^{k-1}, p^k$ to a_1. (This is just another way of saying that the cycle consists of the orbit of a_1 under p). It follows that p^k maps a_1 back to itself, but all smaller powers don't. Since the same cycle can be written as $(a_2\ a_3\ \ldots\ a_k\ a_1)$, the same argument applies to a_2 and, similarly, to all the other numbers appearing in the cycle. Thus $p^k = e$ and it is the smallest positive such power, that is, p has the order k.

$\square$

We will give a complete determination of the orders of all elements in $\mathcal{Z}_n$ and in S_n later on in this chapter.

Proposition 3.9 *Suppose an element a in a group G has a finite order n, and let m be a positive integer. Then $a^m = e$ if and only if n divides m.*

Proof. We assume $a^m = e$ and show that n divides m (the other direction is easy and is left to the reader). The idea in such cases is always the same: Do division with remainder, and prove that the remainder must be 0. Indeed, if $m = qn + r$ with $0 \le r < n$, then we have $r = m - qn$ and $a^r = a^{m-qn} = a^m(a^n)^{-q} = e$ (why?). Thus we must conclude that $r = 0$, since $0 < r < n$ would contradict the minimality property of the order n.

$\square$

We can now deliver on our earlier promise to determine the order of all elements of S_n. Note that this is a generalization of Proposition 3.8.

Proposition 3.10 *The order of a permutation (given in its disjoint cycle representation) is the least common multiple of the lengths of its cycles.*

Proof. We start with the observation that, for disjoint cycles p, q we have $(pq)^n = p^n q^n$. This follows from the fact that disjoint cycles commute. Now if l is the least common multiple of the lengths of p and q, then it follows from Proposition 3.8, p, 97 that $p^l = e$ and $q^l = e$ so, by the opening observation, $(pq)^l = e$. It is left to show that l is the smallest such integer and that the same argument can be extended to the product of any number of cycles. We leave this as an exercise.

$\Box$

We will begin our study of generators by considering the set of all powers (positive, negative and zero) of a single element.

Remark. It is easy to see that the set of all powers of an element forms a subgroup. Moreover, any subgroup of G which contains the element g must also contain all powers of g. In this sense, the set of powers of g is the minimal subgroup of G containing g.

Definition 3.6 (subgroup generated by an element.) *Let g be an element of a group G. The set of all powers (positive, negative and zero) of g is called the* subgroup generated by g. *We denote it by $\langle g \rangle$.*

The following theorem gives a complete description of $\langle g \rangle$ in both the finite and infinite cases.

Theorem 3.2 (what $\langle g \rangle$ really looks like.)

1. *If the order of g is infinite, then $\langle g \rangle = \{g^n : n \in \mathcal{Z}\}$ and the powers are all different ($n \neq m \Rightarrow g^n \neq g^m$). Moreover, the product of elements in $\langle g \rangle$ satisfies the following rule:*

$$g^n g^m = g^{m+n}$$

All this is summarized by saying that $\langle g \rangle$ is isomorphic to $\mathcal{Z}$.

2. *If the order of g is a finite number k, then $\langle g \rangle = \{e, g, g^2, \ldots, g^{k-1}\}$ and these powers are all different. Moreover, the product of elements in $\langle g \rangle$ satisfies the following rule:*

$$g^n g^m = g^{(m+n) \bmod k}$$

All this is summarized by saying that $\langle g \rangle$ is isomorphic to $\mathcal{Z}_k$.

Proof. We leave the proof of 1. for the exercises. For 2., we observe that the main point of the proof is that in the sequence $e, g, g^2, g^3, \ldots$, the first repetition must be $g^k = e$. In fact, suppose that g^m is the first repetition, say $g^m = g^j$, $0 \leq j < m$. "Dividing" both sides by g^j (i.e., multiplying by g^{-j}) we get $g^{m-j} = e$. Since g^m is the first repetition and $0 < m - j \leq m$, it must be that $j = 0$ and therefore $g^m = g^j = e$. (Note the similarity with our discussion of orbits, p. 68.) We leave the remaining details for the exercises.

Theorem 3.2 can be interpreted as follows.

> *Every cyclic group (See Section 3.2.3) is either $\mathcal{Z}$ or one of $\mathcal{Z}_n$, $n = 1, 2 \ldots$*

By this we mean that except for the names of the elements (i.e. replacing g^k by k), the group is the same as $\mathcal{Z}$ or $\mathcal{Z}_n$, depending on its order. In fact, using the identification $k \leftrightarrow g^k$, we can translate statements back and forth between $\langle g \rangle$ and the corresponding $\mathcal{Z}$ or $\mathcal{Z}_n$, in such a way that true statements in one group remain true in the other. In this sense we can say that the two groups have precisely "the same" properties.

Later, when we want to be more precise about two groups being "the same" in this sense, we will introduce and use the term "isomorphic". For now, we explain what this means in the context of cyclic groups.

We will say that a group G is *isomorphic to $\mathcal{Z}$* if it is cyclic, of infinite order and, given the generator g of G, it satisfies

$$g^k \cdot g^j = g^{k+j}.$$

We note that this last condition will always be satisfied by elements of $\mathcal{Z}$.

We will say that a group G is *isomorphic to $\mathcal{Z}_n, n = 1, 2, \ldots$* if it is cyclic, of order n and, given the generator g of G, it satisfies

$$g^k \cdot g^j = g^{(k+j) \bmod n}.$$

We note that this last condition will always be satisfied by elements of $\mathcal{Z}_n$.

You should keep these remarks in mind when we go back and forth in our discussion of cyclic groups between considering $\mathcal{Z}$ or one of $\mathcal{Z}_n$ on one hand, and an "abstract" cyclic group $\langle g \rangle$ of infinite or finite order on the other.

Remark: not all powers are necessary. In view of the previous theorem it should be clear that if the order of g is finite, say k, then all of the powers, including positive, negative and zero, are included in the first k powers of g.

Corollary 3.1 (the two meanings of order.) *The order of an element is the same as the order of the subgroup it generates. In symbols we can write* $order(g) = order(\langle g \rangle)$.

3.2.3 Cyclic groups

The previous discussion suggests the consideration of groups generated by a single element as a separate class. Unlike the general situation, it turns out that it is possible to make a complete analysis of the subgroup structure of this class of groups.

Definition 3.7 (cyclic group.) *A group G is called a* cyclic group *if it has an element g such that $G = \langle g \rangle$. The element g is then called a* generator *of G.*

Examples of cyclic groups. $\mathcal{Z}$ and all the $\mathcal{Z}_n$, $n = 1, 2, \ldots$ are cyclic because they are each generated by 1. The subgroup of rotations in D_4 is generated by R_1 (rotation by $90°$). The group $R = \{1, -1\}$ is generated by -1. The subgroup $\{1, -1, i, -i\}$ is generated by i.

Examples of non-cyclic groups. The Klein group K, S_3 and S_4 are not cyclic groups.

Not only can we determine all cyclic groups as being either $\mathcal{Z}$ or one of $\mathcal{Z}_n$ as we saw in the last section, but we can also give rather complete information on subgroups of any cyclic group. In this section we will establish that every subgroup of a cyclic group is cyclic and hence again is either $\mathcal{Z}$ or one of $\mathcal{Z}_n$. More specifically, every subgroup of $\mathcal{Z}$ other than $\{0\}$ is isomorphic to $\mathcal{Z}$ and in any $\mathcal{Z}_n$ there is exactly one subgroup isomorphic to $\mathcal{Z}_d$ for each divisor d of n. You probably have noticed some of this in Activities 4, 5, and 7. We will state this theorem in full, explain why it is important, discuss the idea of its proof and then give a detailed proof.

Theorem 3.3

1. *Every subgroup of a cyclic group is cyclic.*

2. *For each non-zero integer n the set $n\mathcal{Z}$ of all multiples of n is a subgroup of $\mathcal{Z}$ isomorphic to $\mathcal{Z}$.*

3. *The subgroups $n\mathcal{Z}$, $n = 0, 1, 2, 3, \ldots$, are all the subgroups of $\mathcal{Z}$. Thus, every subgroup of $\mathcal{Z}$ (except $\{0\}$) is isomorphic to $\mathcal{Z}$.*

4. *Let n be a positive integer, and let k, d be divisors of n such that $n = kd$. Then the set $k\mathcal{Z}_n$ of all multiples (mod n) of k by elements of $\mathcal{Z}_n$ is a subgroup of $\mathcal{Z}_n$ isomorphic to $\mathcal{Z}_d$.*

5. *The subgroups $k\mathcal{Z}_n$, k a divisor of n, are all the subgroups of $\mathcal{Z}_n$. Thus every subgroup of $\mathcal{Z}_n$ (except $\{0\}$) is isomorphic to $\mathcal{Z}_d$ for some divisor d of n.*

Remark. This theorem is important because it tells us so much about the subgroups of cyclic groups. The first point asserts that if you are working with a cyclic group, then by looking at subgroups, you do not leave the domain of cyclic groups.

Point 2 of the theorem is very simple. It does not say anything more than that the multiples of a single non-zero element form a subgroup isomorphic to $\mathcal{Z}$. This is just the "what $\langle g \rangle$ really looks like" theorem, expressed in terms of isomorphism. Point 3, however, says much more. It asserts that forming multiples of a single element is the *only* way of getting subgroups of $\mathcal{Z}$. In other words, if you construct all of the subgroups you can by taking multiples of a single element, you have them all and they are all isomorphic to $\mathcal{Z}$.

The situation with cyclic groups of finite order is analogous. Point 4 again repeats the fact that the multiples of an element form a subgroup, but it also contains some information about order. Point 5 tells us that for each decomposition $n = kd$ of n we can find a single subgroup of order d by taking the multiples of k.

The idea of the proof

The proof is based on the following "principle".

A subgroup $H \neq \{e\}$ of a cyclic group G is generated by its "smallest positive" element.

If G is $\mathcal{Z}$ or one of $\mathcal{Z}_n$ then "smallest positive" has its usual meaning. In general, if $G = \langle g \rangle = \{e, g, g^2, g^3, \ldots\}$, we use the identification $k \leftrightarrow g^k$ to give meaning in G to the terms "smallest" and "positive". In this case then, our "principle" reads:

H is generated by g^k where k is the smallest positive exponent of g that appears in the elements of H.

Remark. If a is any non-identity element in H, then either a or a^{-1} is "positive". Thus any $H \neq \{e\}$ has "positive" elements, hence a smallest one.

Proof of the "principle". Suppose that H is a subgroup of the cyclic group $G = \langle g \rangle$. Suppose also that g^k is the "smallest positive" power of g in H; that is, $g^k \in H$ and if $0 < j < k$ then $g^j \notin H$. Let us denote g^k by h. We have to show that under these conditions, $H = \langle h \rangle$.

We assume then that g^m is in H and want to show that g^m is a power of k, that is, $k | m$. The standard way to prove such claims is by showing that the remainder of m on division by k is 0. We can write $m = qk + r$, where the remainder r satisfies $0 \leq r < k$. Since $r = m - qk$, we have

$$g^r = g^{m-qk} = ((g^m)(g^k)^q)^{-1},$$

which is also in H because g^m and g^k are in H and H is a subgroup. Since $g^r \in H$ with $0 \leq r < k$, it follows from the minimality of k that r must be 0, which is the desired conclusion.

$\square$

With this "principle" as a foundation, the proof of the theorem is reduced to checking simple details.

Proof of theorem.

1. For a subgroup $H \neq \{e\}$, this is what the "principle" tells us. The subgroup $\{e\}$ is obviously cyclic.

2. In $\mathcal{Z}$ the multiples of a single element are just the powers of that element in the sense of the group operation. Thus $n\mathcal{Z}$ is the group generated by n. This is a cyclic group by definition. It is also of infinite order, since the multiples of a non-zero integer form an infinite set. Thus it is isomorphic to $\mathcal{Z}$.

3. Suppose that H is a proper subgroup of $\mathcal{Z}$. By the above remark it has a smallest positive element m. By the "principle", H is the subgroup generated by m. By point 2, H is isomorphic to $\mathcal{Z}$.

4. Suppose that $n = kd$. The multiples of k are the powers in the sense of the group operation, so $k\mathcal{Z}_n$ is a cyclic subgroup of $\mathcal{Z}_n$.

   From simple arithmetic we can see that

   $$k\mathcal{Z}_n = \{k, 2k, 3k, \ldots, (d-1)k, 0\}$$

   where the last term comes from the fact that $dk = n$ and $n = 0 \bmod n$. Hence $k\mathcal{Z}_n$ is isomorphic to $\mathcal{Z}_d$.

5. This follows from the "principle" and 4: If H is a proper subgroup of $\mathcal{Z}_n$, then it is generated by its smallest positive element, k. Also, we have seen in the proof of the "principle" that if $g^i \in H$ then $k|i$. Interpreting this for $\mathcal{Z}_n$, using the generator 1 and the operation of addition mod n, this statement reads:

   $$\text{if } i \bmod n \in H, \text{ then } k|i.$$

   Since $n \bmod n = 0$, which is certainly in H, it follows that $k|n$. That is, $dk = n$.

$\square$

Remark. One consequence of point 5 of the theorem is that the order of any subgroup of $\mathcal{Z}_n$ divides n, the order of $\mathcal{Z}_n$. We will see at the end of the chapter that this relationship between orders is very general.

We can now deliver on our earlier promise to determine the order of all elements of $\mathcal{Z}_n$. Note that this is a generalization of Proposition 3.8.

Proposition 3.11 *Let a be any element in $\mathcal{Z}_n$ and let d be the greatest common divisor (g.c.d) of a and n. Then the order of a in $\mathcal{Z}_n$ is n/d.*

Proof. By Corollary 3.1, it is enough to show that $\langle a \rangle = \langle d \rangle$. Since a is a multiple of d, clearly $\langle a \rangle \subset \langle d \rangle$. On the other hand, it is a fundamental property of the integers that since d is the g.c.d. of a and n, we can express d as a "linear combination" of a and n: $d = xa + yn$. Taking this equality mod n, we see that d is in $\langle a \rangle$, hence the reverse inclusion $\langle d \rangle \subset \langle a \rangle$ also holds.

$$\square$$

3.2.4 Generators

We will consider an arbitrary set S of elements in a group G. We will not distinguish between finite and infinite order so we will have to be specific about inverses. We denote by $\langle S \rangle$ the set of all products of finitely many elements of G, each of which is either an element of S or the inverse of an element of S.

Proposition 3.12 (group generated by a set.) *If S is a subset of a group, then $\langle S \rangle$ is a subgroup. This subgroup is the smallest subgroup containing S.*

Proof. Exercise 9.

Note again that by "smallest" we mean here that any subgroup of G which contains S must also contain $\langle S \rangle$.

Definition 3.8 (generators of a group.) *If S is a subset of a group then $\langle S \rangle$ is called the* subgroup generated by S. *If $\langle S \rangle = G$ then S is called a* set of generators *of G. Alternatively, we say that G is generated by S.*

Generators of S_n

We will discuss three different sets of generators of S_n: all cycles, all transpositions (cycles of length 2), and a set with only two elements—a cycle of length n and a transposition. The first set of generators is an immediate consequence of the cycle decomposition theorem, p. 91.

Proposition 3.13 *S_n is generated by all cycles in it.*

Next we show that a very restricted subset of cycles actually suffices.

Proposition 3.14 S_n *is generated by all transpositions in it.*

Proof. S_n is generated by its cycles so it is enough to show that every cycle can be written as a product of transpositions (Why?). To do this you need only check

$$(a_1 \, a_2 \, a_3 \, \ldots \, a_{k-1} \, a_k) = (a_1 \, a_k)(a_1 \, a_{k-1}) \ldots (a_1 \, a_3)(a_1 \, a_2) \, .$$

$\Box$

Note that in one sense this seems a stronger proposition than the cycle decomposition theorem since we show that a smaller set is enough. However, the real strength of the cycle decomposition theorem comes from the facts that the cycles are disjoint and the decomposition is (essentially) unique. None of these properties hold in the decomposition into transpositions.

In the previous sets of generators, the number of generators increases with n. This is only to be expected. It might therefore come as a surprise that we can also generate S_n (for every n) by a fixed number of generators, in fact two.

Proposition 3.15 S_n *can be generated by two elements—a full cycle and a transposition.*

For example, $(1 \, 2 \, 3 \, \ldots \, n)$ and $(1 \, 2)$. Indeed, we can get all transpositions by successive conjugations of $(1 \, 2)$ by $(1 \, 2 \, 3 \, \ldots \, n)$ (see Proposition 3.7, p. 92), and then use the previous set of generators.

Parity—even and odd permutations

We now look more closely at the second set of generators, that is, transpositions. We know that every permutation can be written as a product of transpositions and we use this to classify all permutations into two kinds.

Definition 3.9 (even and odd.) *A permutation is called* even *(resp. odd) if it can be written as a product of an even (resp. odd) number of transpositions.*

Suppose that we could find a permutation that is both even and odd. Then this definition would be totally useless. The fact that this cannot happen is not at all trivial and reveals an important invariant of such decompositions—*parity*. This is the content of the following theorem.

Theorem 3.4 (even and odd.) *Let p be a permutation. Then every decomposition of p into a product of transpositions has the same parity; that is, either all such decompositions of p have an even number of transpositions, or they all have an odd number of transpositions.*

Proof. The proof is sketched in the exercises.

The most interesting property of this classification is that even and odd permutations behave with respect to permutation product the same way that even and odd numbers do with respect to number addition. This is the content of the next proposition.

Proposition 3.16 (parity of the product of permutations.) *For permutations we have:*

> *even * even is even*
> *odd * odd is even*
> *even * odd and odd * even are odd*

Proof. If p and q are even, we can decompose each of them as a product of an even number of transpositions. But then pq has an even decomposition, namely, first write all the transpositions of p then those of q. The other cases are proved similarly.

$$\square$$

Remark. Can you now see that the similarity with even and odd numbers is more than just superficial? Can you formulate the similarity between the two classifications into even and odd in terms of isomorphism of suitable groups?

Theorem 3.5 *The set of all even permutations in S_n forms a subgroup of S_n. This group is called the* alternating group *on $\{1, 2, ..., n\}$ and is denoted by A_n.*

Proof. Exercise 16.

Determining the parity of a permutation.

We need only collect the individual pieces of information we already have. First, what is the parity of a cycle? When we showed that transpositions generate S_n, we saw that a cycle of length k can be decomposed as a product of $k - 1$ transpositions. Thus the parity of a cycle is the opposite of the parity of its length. That is, if it has an even length then it is an odd permutation and vice versa. For example, a cycle of length 3 is even while a cycle of length 4 is odd. The parity of any permutation (written as a product of disjoint cycles) can now be determined from the theorem on the parity of a product. For example, the permutation (1 2)(3 4 5)(6 7 8)(9 10) is even (odd * even *even * odd is even).

3.2.5 EXERCISES

1. Prove that if $a \in \mathbb{Z}_n$ divides n, then the order of a is $\frac{n}{a}$ (Proposition 3.8a).

2. Let $S_{\mathcal{Z}}$ be the set of all permutations of $\mathcal{Z}$. Find $p, q \in S_{\mathcal{Z}}$ such that order$(p) = $ order$(q) = 2$, but pq has infinite order.

3. Show that the "quadratic equation" $x^2 = e$ has an infinite number of solutions in $S_{\mathcal{Z}}$.

4. Show that the union of all S_n, $n = 2, 3, \ldots$ can be identified as the set of all permutations of $\mathcal{Z}$ with finite support.

5. Prove Theorem 3.2(1).

6. Complete the details in the proof of Theorem 3.2(2).

7. Prove Proposition 3.13.

8. Show that the subgroup $\langle 4, 6 \rangle$ generated by $\{4, 6\}$ in $\mathcal{Z}$ is the same as the subgroup $\langle 2 \rangle$ generated by $\{2\}$. Also show that $\langle 3, 5 \rangle = \langle 1 \rangle = \mathcal{Z}$.

9. Let S be a subset of the group G. Prove the following relation,

$$\langle S \rangle = \bigcap \{H : S \subset H \text{ and } H \text{ is a subgroup of } G\}$$

Deduce from this a proof of **Proposition 3.12**.

10. List all subgroups of $\mathcal{Z}_{30}$ and their orders.

11. Find the orders of all elements of $\mathcal{Z}_{30}$.

12. List all generators of $\mathcal{Z}_{30}$.

13. Prove in detail that S_n is generated by $(1\ 2\ 3\ \ldots\ n)$ and $(1\ 2)$ (Proposition 3.15).

14. Prove the "push left" Lemma: If $t_1 = (xy)$ and $t_2 = (ab)$ are disjoint transpositions, then there are transpositions t_1' and t_2' consisting of the same numbers x, y, a, b such that $t_1 t_2 = t_1' t_2'$ and where a appears only in t_1'. (The name is meant to invoke the image of "pushing" a one place to the left without affecting the value of the product.)

15. Prove that all decompositions of a given permutation, $p \in S_n$, into a product of transpositions have the same parity (Theorem 3.4), that is, either they are all even or they are all odd. (We will refer to such decompositions as even or odd, respectively.)

    Guidance: (You are asked to fill in the gaps in the following outline.)

    (a) The proof follows easily from the following two main ideas:
    One, it is enough to prove that all decompositions of the *identity* permutation I as a product of transpositions are even. Two, if I

can be written as a product of r transpositions with $r > 2$, then I can also be written as a product of $r - 2$ transpositions.

Assuming that these two claims have been proved, show how our theorem follows from them.

(b) Proof of the first main idea: Show that if a certain permutation has both an even and an odd decomposition as a product of transpositions, then you can get an odd decomposition of the identity permutation I. Conclude that it is enough to prove that all the decompositions of the identity permutation are even.

(c) Proof of the second main idea: You need to show that if I can be written as a product of r transpositions with $r > 2$, then I can also be written as a product of $r - 2$ transpositions. The idea here is that subproducts of the form $(ab)(ab)$ can simply be eliminated from a product of transpositions without affecting its value. Suppose then that for some $r > 2$ we have a decomposition

$$I = t_1 t_2 \cdots t_{r-1} t_r$$

and let a be a number appearing in t_r. If $t_r = t_{r-1}$ then we are finished. If not, we use the "push left" Lemma (previous exercise) to successively push a to the left until we get a decomposition of I containing a segment of the form $(ab)(ab)$. This must happen or else we would eventually get a decomposition of I in which a apears only in the left-most transposition, which is impossible.

16. Prove that A_n, the set of all even permutations in S_n, is a subgroup of S_n (Theorem 3.5). Is the set of all *odd* permutations a subgroup?

17. How many even and how many odd permutations are there in S_n?

18. Let H be a subgroup of S_n which is not a subgroup of A_n. Prove that H has the same number of even and odd elements.

19. Show that the parity of a permutation is the same as the parity of the number of its cycles which are odd permutations.

20. Here is a neat graphical trick for determining the parity of a permutation. Explain why it works.

In the standard two-row notation for a permutation, locate the two occurences of 1 in the two rows and draw the line segment connecting them. Do the same for all other numbers involved. Take care not to have three line segments intersecting at the same point. Now count the number of points of intersection of pairs of these line segments. The parity of the permutation is the same as the parity of the number of intersection points.

For example, the permutation $\begin{pmatrix} 1\,2\,3\,4 \\ 2\,4\,3\,1 \end{pmatrix}$ will give rise to four inter-section points, hence it is even. (Note: Had we considered the above permutation as a permutation on the set 1,2,4 only, it would have had only 2 intersection points, but the parity is still even. This kind of invariance holds in general.)

21. Are there any non-cyclic groups of order 5?

22. What happens in $\mathcal{Z}_n$ with the multiples of a number k which is neither a factor nor relatively prime to n?

23. Show that S_{20} has a cyclic subgroup of order 99.

24. Assume that n, a, b are positive integers such that $n = a + b$ and a, b are relatively prime (that is, have no common divisor greater than 1). Show that S_n has a cyclic subgroup of order ab.

25. What is the maximal order of an element of S_{10}?

26. Show that a 3-cycle is an even permutation and that for $n \geq 3$, A_n is generated by the set of all 3-cycles in it.

27. Prove or give a counter-example to the following statement: A group, all of whose proper subgroups are cyclic, is itself cyclic.

28. Give two different proofs of the assertion that conjugate permutations have the same parity.

3.3   Lagrange's theorem

3.3.1   ACTIVITIES

1. Given a group G with operation o, we can form the *generalized group product* oo as follows: The generalized product accepts two inputs which can be any combination of elements of G and/or subsets of G. If the two inputs are both elements, say a and b, then $a\ oo\ b$ is the same as $a\ o\ b$. If the first is an element a and the second is a subset S then $a\ oo\ S$ is the set of all products $a\ o\ b$, $b \in S$. Similarly for the other two cases.

Write an **ISETL func** PR that, assuming **name_group** or **group_table** has been run, forms the generalized product. That is, PR accepts two inputs, a group and an operation, and returns a **func** which represents the generalized product. Thus, if you write

$$oo := PR(Z12, a12)$$

then you will have defined oo as the generalized product for that group.

Run your func in the following situations. You can use oo to denote the generalized operation which is returned by PR as in g .oo S or S .oo T.

(a) G = Z20, inputs are 11, 19.

(b) G = S4, inputs are (1 2 3), (1 2)(3 4).

(c) G = S4, inputs are (1 2)(3 4), (1 2 3).

(d) G = Z20 inputs are 11, {0, 5, 10, 15}.

(e) G = Z20 inputs are {0, 5, 10, 15}, 11.

(f) G = Z24 inputs are {0, 8, 16}, {0, 5, 10, 15}.

(g) G = S3, inputs are {(1), (1 2 3), (1 3 2)}, (1 2).

(h) Same as previous one with the inputs in reverse order.

(i) G = S3, inputs are (1 2 3), {(1), (1 2 3), (1 3 2)}.

2. In this activity you are to use PR to investigate a certain situation in the group S4.

Your task is to investigate a partial commutativity condition for .oo. That is, if H is one of the two subgroups K1 or K2 (see p. 88), is it the case that for every x in G, H .oo x = x .oo H ? Is this always the case? Sometimes? Never? Is your answer the same for S = K1 as for S = K2?

3. In this activity, you are going to use PR again for a group G with input pairs consisting of an element and a fixed subgroup H, and the task is to run through all elements x of G. Thus your result will be a set of subsets of G. These are the sets H .oo x, x in G.

The questions we ask are: How many elements in each of the subsets? How many subsets are there? Which elements of G are in two or more of these subsets? Which elements are not in any of them?

Calculate the set of subsets and answer the questions in each of the following situations.

(a) G = Z24, H is the subgroup generated by 6.

(b) G = S4 and H is the subgroup K1 of Activity 2.

(c) G = S4 and H is the subgroup K2 of Activity 2.

4. After completing Activity 3, you should be able to summarize your observations and make a conjecture about what happens in general. Do so.

5. The objects you worked with in the previous two activities are called cosets. x .oo H is called a left coset of H and H .oo x is called a right coset of H (we will give a more formal definition in the text). Write **ISETL** funcs left_coset and right_coset. They will accept an element of G and a subgroup H and return the corresponding left, resp. right coset of H.

   Select five examples (with different groups, subgroups and elements) from the previous activities or make up your own and compute left and right cosets in each case.

6. For various examples of groups G and subgroups H, construct the set GmodH of all left cosets of H with elements of G, and investigate its properties (e.g., how many elements does it have, how many elements does each of its elements have, etc.).

7. In Activity 1 of Section 2.3.1, p. 71, you defined the proc name_ group that assigns to a given group the standard names G,o,e and i. Following is an enhancement to this proc that gives standard names also to a subgroup and its related objects. The inputs to this new version of **name_group** are a set, a binary operation and an additional *optional* input which is a subset of the original set, assumed to be a subgroup. (Note the new keyword opt preceding the optional input parameter subst.) The new proc can be executed either with the first two inputs (set and operation) only, or with all three inputs. In the first case the result will be just as in the old **name_group** (except for the addition of oo), but in the second case all the following names will be assigned:

   G - the group.
   o - the group operation.
   e - the identity of G.
   i - the func that for each element g in G returns its inverse.
   oo - the generalized group operation.
   H - the subgroup.
   K - the func that for each g in G returns the right coset H .oo g.
   GmodH - the set of all right cosets H .oo g for g in G.

   You can enter !pp name_group (do not end with ;) to get the text of the old definition of **name_group** on the screen, then make the neccessary changes to define the new version.

```
name_group := proc(set,operation opt subst);
        G := set; o := operation; oo := PR(G,o);
        e := identity(G,o); i := |g -> inverse(G,o,g)|;
        writeln "Group objects defined: G, o, oo, e,
        i .";
        if is_defined(subst) then
            H := subst;
            GmodH := {H .oo g : g in G}; $ The sets of
            right cosets
            K := | g -> H .oo g |; $ K(g) is the coset
            of g
            writeln "Subgroup objects defined: H, GmodH,
            K .";
        end;
    end;
```

Run the new **name_group** on various examples, with or without the optional subgroup, and make sure you understand what all the standard names in the above list mean. Save this **proc** in some file as you will be using it in the future.

8. Let G be a group and H a subgroup. In this activity you are to form the system consisting of the set GmodH and the binary operation .oo on it. Run is_group to see if this system forms a group in the following situations.

   (a) G = Z24, H is the subgroup generated by 6.

   (b) G = Z15, H is the subgroup generated by 5.

   (c) G = S4 and H is K1.

   (d) G = S4 and H is K2.

3.3.2   What Lagrange's theorem is all about

Lagrange's theorem is one of the most important theorems of elementary group theory. It has many consequences, both general and particular. The theorem tells us a great deal about what groups are possible. For example, you will see that every group of order 5 is cyclic, hence isomorphic to $\mathcal{Z}_5$. The task of finding all groups of some other orders such as 6, 7 and 8 is greatly simplified by using Lagrange's theorem. It also tells us a great deal about subgroups of a given group. For example, S_3 cannot have any subgroup of order 4, S_4 cannot have any subgroups isomorphic to $\mathcal{Z}_5$ and $\mathcal{Z}_{17}$ cannot have any (proper) subgroups at all! Finally, there are a number of theoretical consequences of major importance in group theory.

We will first consider the theorem and some of its consequences. Then, in the next section, we will look into the proof.

Theorem 3.6 (Lagrange's theorem.) *The order of the subgroup of any finite group divides the order of the group.*

Remark. Now you can see why S_4 cannot have a subgroup of order 5 — simply because 5 does not divide 24. Also, since 17 is a prime, its only divisors are 1 and 17, so the only subgroups $\mathcal{Z}_{17}$ can have must be of order 1 and 17. Thus they must be either $\{0\}$ or $\mathcal{Z}_{17}$. All this has the following generalization.

Corollary 3.2 *A group of prime order has no proper subgroups.*

Proof. Exercise 1(a).

Remark. Look again at $\mathcal{Z}_{17}$ as an example of a group of prime order. If $k \in \mathcal{Z}_{17}$ and $k \neq 0$, then since the order of k divides 17 it must be 17. This means that $\langle k \rangle = \mathcal{Z}_{17}$. Again we have the generalization.

Corollary 3.3 *A group of prime order is cyclic. Moreover, it is generated by any of its elements other than e.*

Proof. Exercise 1(b).

Following are two straightforward consequences of Lagrange's theorem.

Corollary 3.4 *The order of an element in a finite group divides the order of the group.*

Proof. See Corollary 3.1, p. 99.

Corollary 3.5 *In a group of order n, every element g satisfies the equation* $g^n = e$.

Proof. Exercise 3.

3.3.3   Cosets

You have worked extensively in this chapter with cosets. By now the ideas should be solidly established in your mind so we are ready to formalize them. Although we often speak loosely about "cosets" and this makes sense in a commutative group, in general, there are really two kinds of cosets — left cosets and right cosets. The main ingredients in a coset are a subgroup of a group and an element of the group.

Definition 3.10 (cosets.) *Let G be a group, H a subgroup, and $a \in G$. The right coset of H by a is the set Ha given by*

$$Ha = \{ha : h \in H\}.$$

Similarly, the left *coset of H by a is the set aH given by*

$$aH = \{ah : h \in H\}.$$

If G is commutative, then we simply say the coset of H by a.

The individual properties of cosets are simple but important, and with them, cosets form a powerful tool in the theory of groups. In the following proposition, we state some of these properties for right cosets and leave it to the reader to formulate the analogues for left cosets. You have actually dealt with these properties in the activities. Hence, Proposition 3.17 should be a summary and formalization of facts that you already know.

Proposition 3.17 *Let H be a subgroup of G and $a, b \in G$. Then the following statements hold.*

1. $a \in Ha$.

2. $Ha = H$ *if and only if* $a \in H$.

3. $Ha = Hb$ *if and only if* $ba^{-1} \in H$.

4. *The orders of Ha, Hb are equal.*

5. *Either $Ha = Hb$ or $Ha \cap Hb$ is empty.*

Proof. Exercise 8.

3.3.4   The proof of Lagrange's theorem

The formal proof is rather short and apparently simple, but our experience has been that many students are mystified by it on first reading. We'd therefore like to "unpack" it by first describing informally the *idea* of the proof and by looking at an example.

The proof of Lagrange's theorem makes use of cosets. It uses them in a construction that is analogous to what we did with orbits and cycles (Chapter 2, Section 2, Activities 8, 9, p. 58 and the discussion preceeding Theorem 3.1, p. 91).

The idea goes something like this. Recall our initial discussion of orbits. We took a permutation of $\{1, 2, ..., n\}$ and applied it successively to a single number. This produced a subset of $\{1, 2, ..., n\}$. We then took one of the numbers that had not been included and repeated the process. This gave another subset which included none of the elements from the previous subset. We continued this until all of $\{1, 2, ..., n\}$ was used up. We did the same thing in decomposing a permutation into cycles. In each case the idea is that we keep generating subsets of $\{1, 2, ..., n\}$, never repeating a number, until all of $\{1, 2, ..., n\}$ is used up.

We do the same thing with cosets of a subgroup to prove Lagrange's theorem. The facts are a little harder to prove and there is an additional twist that is essential for the proof, but the basic idea is the same. We have a group G and a subgroup H. We pick an element g of G and form the right coset Hg by forming every product $hg, h \in H$. This gives a subset of G. We pick an element of G that is not in the subset and form its right coset. We get a new subset and we have to prove that we don't repeat any of the elements of the first subset. This gives another subset and we keep doing this until all of G is used up.

Thus again we get a collection of subsets of G with every element appearing in one and only one subset. The added twist is the fact (we have to prove it) that all of these subsets have the same number of elements.

Now think about it. We have a decomposition of G into cosets, each having the same number of elements, and so the order of G must be the number of cosets times the number of elements in a coset. Hence the number of elements in a coset divides the order of G.

Finally we note that when we picked an element of H to form the coset, we get H. That is, H is one of these cosets. Therefore, the number of elements in H divides the order of G.

Example. Before we formalize this proof, let's take a look at a typical example that you worked with in Activities 3 and 6, p. 109ff. Take the group $G = S_4$ and let $H = \{p \in S_4 : p(4) = 4\}$. You can use your coset func or your mind to figure out the 4 right cosets of H. For instance if you take the permutation (3 2 1 4), then the coset $H(3\,2\,1\,4)$ you get is

$$L_1 = \{(3\,2\,1\,4),\ (2\,3\,1\,4),\ (3\,1\,4),\ (2\,1\,4),\ (2\,3)(1\,4),\ (1\,4)\}$$

and if you take (1 3)(2 4), you get

$$L_2 = \{(1\,3\,2\,4),\ (3\,1\,2\,4),\ (1\,2\,4),\ (3\,2\,4),\ (1\,3)(2\,4),\ (2\,4)\}.$$

The other two are

$$L_3 = \{(1\,2\,3\,4),\ (2\,1\,3\,4),\ (1\,3\,4),\ (2\,3\,4),\ (1\,2)(3\,4),\ (3\,4)\}.$$

$$L_4 = He = H$$

Note that (3 2 1 4) is in $H(3\,2\,1\,4)$ because you must include $(3\,2\,1\,4)e$. This is what always happens. Also, you get 6 elements every time because you must multiply g by all 6 elements of H, and moreover, there are no repetitions. Finally, every element is eventually taken because you keep doing this until the whole group is used up.

Now let's turn to the formalization. The only point that is not really clear from the examples is why two different cosets never have an element in common. This will follow immediately from the following lemma.

Lemma 3.1 *Let H be a subgroup of G and $a, b \in G$. If $b \in Ha$ then $Ha = Hb$.*

Proof. We have some $h \in H$ with $b = ha$. Then for any $x \in H$, $bx = hax \in H$ which shows that $Hb \subset Ha$. Conversely, if $x \in H$ then $ax = h^{-1}bx \in Hb$ which shows that $Ha \subset Hb$. Hence $Ha = Hb$.

$\Box$

Now we are ready for the formal proof.

Proof of Lagrange's theorem. Let H be a subgroup of the finite group G. Suppose that H has order k. We form the set of all right cosets of H by elements of G which we write as,

$$GmodH = \{Hg : g \in G\}.$$

Clearly every element of G is in one of these sets since $g = eg \in Hg$. Could any element be in two of these sets? That is, could we have some g with $g \in Hg_1$ and $g \in Hg_2$, with $Hg_1 \neq Hg_2$? No, because we would then have, by Lemma 3.1, $g \in Hg_1$ and $g \in Hg_2$, so $Hg_1 = Hg_2$. Hence, no element of G is in two of these cosets.

How many elements are there in one of the cosets Hg? Well, we have

$$Hg = \{hg : h \in H\}$$

so we compute one for each $h \in H$ or k of them. Could we get a repetition? That is, could we have $h_1g = h_2g$? According to the cancellation law, this only happens if $h_1 = h_2$. So we really do get k different elements in Hg. Thus, every coset has k elements.

To summarize, we have the collection of cosets GmodH, every element of G is in one and only one of them and each coset has k elements. Hence the order of G is k times the number of cosets. Thus k divides the order of G.

$\Box$

Applications. Lagrange's theorem and its corollaries makes it reasonable to determine a number of facts about specific groups. For example, it follows immediately from Corollary 3.3, p. 112, that if p is a prime, then there exists exactly one group of order p. Following are some additional facts which are a bit more difficult to prove. In each case, the method of proof consists of considering orders of elements and orders of subgroups. We leave the details for the exercises.

Proposition 3.18 *There are exactly two groups of order 6, one is commutative ($\mathcal{Z}_6$) and the other is not (S_3).*

Proposition 3.19 *The group S_3 has one subgroup of order 1 (the trivial group), three subgroups of order two (each two cycle together with the identity), one group of order 3 (the identity together with the two 3-cycles), and one subgroup of order 6 (the whole group).*

3.3.5 EXERCISES

1. Let G be a group of finite prime order. Prove the following.

   (a) G has no proper subgroups (Corollary 3.2, p. 112).

   (b) G is cyclic, in fact generated by any of its non-identity elements (Corollary 3.3, p. 112).

2. Prove that the order of an element in a finite group divides the order of the group (Corollary 3.4).

3. Prove that in a group of order n every element satisfies the equation $x^n = e$ (Corollary 3.5).

4. Prove Proposition 3.18, p. 115.

5. Prove Proposition 3.19, p. 115.

6. Show that any non-commutative group of order 6 is isomorphic to S_3.

7. Prove that in a finite Abelian group whose order is a prime other than 2, the product of all the group elements is equal to e. Is this also true in non-Abelian groups?

8. Prove Proposition 3.17, p. 113.

9. Prove that if H is a subgroup of G with only two right cosets, then every left coset of H in G is also a right coset. What does this imply about the subgroup A_n of S_n?

10. Let H be a subgroup of any group G (finite or infinite). Establish the existence of one-to-one maps from any right coset Ha onto the corresponding left coset aH, and from the set of all right cosets of H in G onto the set of all its left cosets. Show that in the finite case this implies that the corresponding sets have the same size.

11. Suppose H is a subgroup of a group G, $a, b \in G$. If $Ha = Hb$ we say that a is *equivalent* to b mod H and we write this as $a \equiv b \pmod{H}$.

    (a) Prove that $a \equiv b \pmod{H}$ if and only if $ab^{-1} \in H$.

    (b) Show that this is indeed an equivalence relation, i.e. it is reflexive, symmetric and transitive. (Compare Exercise 8, p. 55 in Section 2.1.5 and Proposition 3.6, p. 90.)

    (c) Show that the equivalence class of $a \in G$ mod H that is, the set of all elements equivalent to a, is precisely the coset Ha.

12. In the case of $G = \mathcal{Z}$ and $H = \langle n \rangle$, show that $a \equiv b \pmod{H}$ has the same meaning as $a \equiv b \pmod{n}$. Determine all the equivalence classes. How many such classes are there and how many elements in each?

13. In the case $G = S_4$ and $H = \{p \in S_4 : p(4) = 4\}$ (recall that H is isomorphic to S_3), show that $q \equiv r \pmod{H}$ if and only if $q(4) = r(4)$. Determine all the equivalence classes. How many such classes are there and how many elements in each?

14. Prove Fermat's Little Theorem: For every integer a and prime p, $a^p \equiv a \pmod{p}$.

    (Hint: This is equivalent to $a^{p-1} \equiv 1 \pmod{p}$ for every *non-zero* integer a. Use the multiplicative group of integers mod p.)

15. Prove that a group which has no non-trivial subgroups must be finite of prime order.

16. Prove that if G is a non-cyclic group of order 49, then all its elements satisfy the equation $x^7 = e$.

17. Let G be the group of all $n \times n$ real matrices with non-zero determinant, and H the subgroup of all matrices $A \in G$ such that $\det A = 1$. If A, B are matrices in G such that $HA = HB$, what can you say about the relationship between $\det A$ and $\det B$? Characterize the cosets of H in G.

4

The Fundamental Homomorphism Theorem

4.1   Quotient groups

4.1.1   ACTIVITIES

1. In the last section of the previous chapter you constructed, given a group G and a subgroup H, the set of cosets GmodH and coset multiplication in **ISETL**. The cosets were always right cosets. In the next few activities you'll be exploring the relationship between right and left cosets. Edit the proc name_group to include the set of right cosets, left cosets and multiplication between cosets. Also, replace K by two funcs, Kr and Kl corresponding to right and left cosets. (You'll need your original name_group later so it might be a good idea to save a copy of it under a different name before editing.) Now after running name_group you should have the following notation.

   G — the group
   o — the group operation
   oo — the generalized product
   e — the identity of G
   i — the func that for each element g in G returns its inverse
   H — the subgroup

Kr — the func giving, for each g in G, the right coset H .oo g
Kl — the func giving, for each g in G, the left coset g .oo H
GrmodH — the set of all right cosets H .oo g for g in G
GlmodH — the set of all left cosets g .oo H for g in G

2. Following is a list of group/subgroup pairs G, H followed by a list of questions. Run your proc from the previous activity on each of the pairs and answer the questions for each pair G, H.

   (a) Z_{12}, $H = \{0, 4, 8\}$

   (b) S_3, A_3

   (c) S_3, $T_1 = \{(1), (1\ 2)\}$

   (d) S_4, A_4

   (e) S_4, $K_1 = \{I, (1\ 2), (3\ 4), (1\ 2)(3\ 4)\}$

   (f) S_4, $K_2 = \{I, (1\ 2)(3\ 4), (1\ 3)(2\ 4), (1\ 4)(2\ 3)\}$

   (g) A_4, K_2

   (h) K_1, $T_2 = \{(1), (1\ 2)\}$

   (i) D_4, the subgroup of all rotations: $\{I, R_1, R_2, R_3\}$

   (j) D_4, the subgroup generated by a reflection: $\{I, V\}$

   Here are the questions.

   i. For which $a \in G$ is it the case that a .oo H = H .oo a?
   ii. For which $a \in G$ is it the case that Kl(a) = Kr(a)?
   iii. Is [GrmodH,oo] a group?
   iv. Is [GlmodH,oo] a group?
   v. Is GlmodH = GrmodH?

3. In Activity 2, for all cases in which GlmodH or GrmodH was not a group, which of the group postulates were not satisfied?

4. In Activity 2, for all cases in which GlmodH or GrmodH was a group, answer the following questions.

   (a) Which is the identity element?

   (b) Specify the inverse of each element.

   (c) Is there any sense in which you can say that the group is "the same" as some group you have previously worked with? (Recall our discussion of isomorphism in Section 2.3.4, p. 76.)

   (d) How does your response to question (c) relate to the discussion in Section 2.3.4?

5. The condition that $Ha = aH$ for every a in G is, in a sense, related to one of the group properties. Which one is it? Can you formulate a precise relation? Can you prove it?

6. In a short while we'll be introducing the term "normal" subgroup for a subgroup satisfying the condition of the last activity: $Ha = aH$ for every $a \in G$. Here is a list of conditions related to coset multiplication. The majority of them are equivalent to the "normality" condition (which appears first on the list) but this does not include all of them. Test these conditions on all of the examples you have studied in this set of activities. Indicate which conditions you can say for sure are **not** equivalent to normality and which, based on the data from these examples, could be equivalent.

   The symbol $\forall$ in the following statements (and elsewhere in this book) is an abbreviation of the phrase "for all".

   (a) $\forall a \in G, \ Ha = aH$.

   (b) $\forall h \in H$ and $\forall a \in G, \ aha^{-1} \in H$.

   (c) $HH = H$.

   (d) $\forall a \in G, \ aHa^{-1}$ is a subset of H.

   (e) $GH = G$.

   (f) $\forall a \in G, \ aHa^{-1} = H$.

   (g) $GH = HG$.

   (h) $HGH = G$.

   (i) $\forall a, b \in G, \ (Ha)(Hb)$ is a right coset of H.

   (j) $\forall a, b \in G, \ (Ha)(Hb) = H(ab)$.

   (k) $GlmodH = GrmodH$

Re-edit your **proc name_group** to include right cosets only, or just retrieve the original version if you saved it. That is, rename GrmodH and Kr as GmodH and K and eliminate GlmodH and Kl.

4.1.2   Normal subgroups

The proof of Lagrange's theorem, which you studied in the previous chapter, hinges on the concept of a coset of a given subgroup. More specifically, for a subgroup H of a finite group G, the proof of Lagrange's theorem utilizes the set of all right cosets of H in G, and the partition it induces on G. In the activities, the set of all right cosets of a subgroup H in a group G was denoted GrmodH, or GmodH for short. We are now going to study this set in more depth. Until further notice, we'll refer to right cosets simply as cosets.

Recall that GmodH has an operation defined on it called the "generalized product" and denoted oo in the activities. If L and M are two cosets of the subgroup H of the group G, then in mathematics we write the product as LM, where

$$LM = \{xy : x \in L, y \in M\}$$

As you saw in Activities 2, 3, 4, and 6(i), it may or may not be the case that LM is again a coset of H. In any case, this is an operation which may or may not satisfy the group postulates. If it does, we call this group the *quotient group of G mod H*. The standard notation in mathematics for the quotient group is G/H, but we'll also retain our ISETL notation GmodH for the set of cosets in cases it is not known to be a group.

Thus, our first goal in studying quotient groups is to determine when the set of cosets is a group under this binary operation, in particular when coset multiplication satisfies the closure postulate. A seemingly more restricted question (which in fact turns out to be equivalent) is, under what conditions is it true that $(Ha)(Hb) = H(ab)$? This condition asserts not only that the product of two cosets (of the elements a, b respectively) is again a coset, but that it is, in fact, the coset of the product ab. In Activity 6, you may have noticed a connection between this condition on the product of cosets and the equality of left and right cosets. This leads to the following fundamental concept of a normal subgroup.

Definition 4.1 (normal subgroup.) *A subgroup N of a group G is said to be* normal *if $Ng = gN$ for all g in G.*

There are many equivalent conditions, besides the one given in the definition, to characterize normal subgroups. We will collect some of them in Proposition 4.2 below. One particularly useful condition, which in fact is taken to be the definition in many textbooks, is the following:

$$\text{For all } x \in N \text{ and for all } g \in G, \quad gxg^{-1} \in N.$$

Another way of saying this is that N is *closed under conjugation*. We leave as an exercise the proof that this condition is indeed equivalent to the one given in the definition.

The condition in the definition may be considered as a weak form of commutativity. In particular, do you see that in an Abelian group all subgroups are normal? Convince yourself, however, that normality is not exactly the same as commutativity.

The importance of normal subgroups stems largely from the following connection with coset multiplication, which hopefully you have already anticipated in the activities.

Proposition 4.1 *If N is a normal subgroup of a group G then $(Ng)(Nh) = N(gh)$ for all g, h in G.*

Proof. We will give two proofs in order to illustrate the connection between the language of individual elements of sets and the generalized product. In both proofs, we begin by assuming that N is a normal subgroup of G.

A typical element of Ngh is of the form xgh for some $x \in N$. Since $e \in N$, writing $xgh = xgeh$ shows that $xgh \in NgNh$. Conversely, consider a typical element $xgyh \in NgNh$, with $g, h \in G$, $x, y \in N$. By our assumption of normality, $gy \in gN = Ng$ and so we have $y' \in N$ with $gy = y'g$. Therefore, $xgyh = xy'gh \in Ngh$ where we have used closure in N to conclude that $xy' \in N$.

The last argument is typical of the way that normality serves as a weak form of commutativity: if we could replace gy by yg (commutativity), we would be done; instead we replace it by $y'g$ for some y' in N.

This completes the first proof. The alternate proof is somewhat simpler using the generalized product notation, the fact that it is associative and that, as for any group, $NN = N$ (exercise). Thus, because of the normality, we can write,

$$(Ng)(Nh) = N(gN)h = N(Ng)h = (NN)gh = Ngh.$$

$\square$

Here are some conditions that are equivalent to normality. Note how several of them can be seen as weak forms of commutativity. We leave the proofs for the exercises.

Proposition 4.2 *For a subgroup H of a group G, the following are equivalent.*

1. *H is normal.*

2. *For all $x \in H, g \in G$, $gxg^{-1} \in H$.*

3. *For all $g \in G$, $gHg^{-1} \subset H$.*

4. *For all $g_1, g_2 \in G$, $Hg_1 Hg_2 = Hg_1 g_2$.*

5. *For all $g \in G$, $gHg^{-1} = H$.*

6. *For all $h \in H, g \in G$, there exists $h' \in H$ such that $hg = gh'$.*

7. *For all $g \in G$, $Hg subset gH$.*

8. *For all $g \in G$, $Hg = gH$.*

9. *$GmodH \subset GlmodH$.*

10. *$GrmodH = GlmodH$.*

In the case of S_n, the group of all permutations of $\{1, ..., n\}$, we have a very useful characterization of normal subgroups. It follows easily from Proposition 3.7, p. 92. First a definition.

Definition 4.2 *Two permutations in S_n, written as a product of disjoint cycles, are said to have the same cycle structure if they have the same number of disjoint cycles, and the respective cycles have the same lengths. A subset of S_n is said to be* closed with respect to cycle structure *if, together with each permutation it contains, it also contains every permutation which has the same cycle structure.*

For example, (1 2 3)(4 5 6)(7 8) and (2 4 7)(6 8)(1 3 5) have the same cycle structure. The subgroup K_2 from Activity 2 is closed with respect to cycle structure, while K_1 is not. Can you see why?

Proposition 4.3 (normal subgroups of $\mathbf{S_n}$) *A subgroup H of S_n is normal if and only if it is closed with respect to cycle structure.*

Proof. It follows from Proposition 3.7 that conjugating an element $a \in S_n$ only permutes its digits, but doesn't change its cycle structure.

Therefore, to say that a subgroup N of S_n is closed under conjugation is the same as saying that N is closed with respect to cycle structure. Since the former statement is equivalent to normality, the result follows.

$\square$

Multiplying cosets by representatives

We elaborate on the meaning of the important Proposition 4.1. Suppose N is a subgroup, K and L are two arbitrary cosets, and we want to multiply them. As a consequence of Lemma 3.1, p. 114, every $a \in K$ gives rise to a representation $K = Na$. For this reason, the elements of K are called *representatives* of K, and similarly for L. Thus, each coset has many representatives. It would simplify calculations with cosets if we could use these representatives, but because the representative is not unique, there is potential for ambiguity. What Proposition 4.1 does is to guarantee that no inconsistency will occur.

Let's be a little more specific. Suppose we tried to calculate the product of two cosets K and L by picking representatives, $a \in K$ and $b \in L$. Then we could form ab and take the coset of that. What relation does this coset have to the product coset KL? That is, if H is the subgroup, what is the relation between $KL = (Ha)(Hb)$ and Hab?

Proposition 4.1 says that if H is normal, then they are the same. What the result prescribes for multiplying cosets of a normal subgroup is the following: choose arbitrary representatives $a \in K$ and $b \in L$; then the product of K and L is the coset represented by ab. Another way of saying this is "to multiply cosets, just multiply their representatives".

Let's consider an example. In Activity 2(a) you worked with the cosets of $H = \{0, 4, 8\}$ in $\mathcal{Z}_{12}$. Let K be the the coset of 3 and L the coset of 9. Thus,

$$K = H +_{12} 3 = \{3, 7, 11\}$$
$$L = H +_{12} 9 = \{1, 5, 9\}.$$

How can we compute the coset "product" $K +_{12} L$? (We use the same notation for the generalized operation and the original group operation.) One way is to use the definition and write

$$K +_{12} L = \{x +_{12} y : x \in K, \ y \in L\}$$

which, after working out all 9 sums (mod 12), turns out to be equal to $\{0, 4, 8\}$, or H. An easier way would be to take representatives — say 7 for K and 9 for L and find the coset of $7 +_{12} 9 = 4$. The coset of 4 is H, so both methods give the same answer. The subtle issue here is the arbitrary choice of the representatives 7 and 9. Could it happen, for example, that by selecting different representatives we would get a different result for the "product" of the same cosets? If this were the case, the product would not really be defined. Let's select different representatives, say 11 for K and 9 for L and compute the product as before. Since $11 +_{12} 9 = 8$, and since $8 \in H$, we indeed get the same coset H as before. Since $\mathcal{Z}_{12}$ is commutative, H is automatically normal. Thus Propositon 4.1 predicts this agreement and tells us that it will always occur for this example, and in general for all normal subgroups.

Our definition of coset multiplication had the advantage that we could work with GmodH and its product long before the introduction of normal subgroups, avoiding the subtle issue of the product being independent of the choice of the representative. At any rate, Proposition 4.1 shows that for normal subgroups, the two definitions are equivalent.

4.1.3   The quotient group

In Activity 2 you considered a number of examples and investigated when a subgroup is normal and when the set of its cosets, with coset multiplication, forms a group. Did you observe that these two phenomena always occur simultaneously? That is, in any example, if one is true, then so is the other. This is always the case and forms the key result of this section.

Remark: The technical details of the proof of the following theorem are very simple. It is important to notice, however, the general idea behind the proof, which is *inheritance*. The new formula for the product given by Proposition 4.1 in fact creates such a close link between the original group G and its "offspring" G/N, that we can use it (the formula) to show how the group properties of G are "inherited" by G/N. Thus the general plan of the proof of each property is as follows: first use the product formula for cosets to translate the statement relative to G/N to a corresponding statement relative to G. Then use the fact that G is a group to conclude that the statement is indeed true in G. Finally, use the product formula again to translate the result back to G/N. We will supply the calculations with little comment. It is important that you try to relate the calculations to the general plan noted here.

Theorem 4.1 (normality and the quotient group.) *If N is a normal subgroup of G, then the set of cosets is a group relative to coset multiplication.*

Proof.

Closure. The formula $(Na)(Nb) = N(ab)$ guarantees that the product of two cosets is indeed a coset.

Associativity. We show how associativity is inherited from associativity in G.

$$(NaNb)Nc = N(ab)Nc = N(ab)c = Na(bc) = NaN(bc) = Na(NbNc)$$

Existence of an identity element. The unit element is the coset Ne, where e is the unit element of G (cosets inherit properties of their representatives). Indeed, for all a in G, $NeNa = Nea = Na$ and similarly $NaNe = Na$. (Note, incidentally, that eN is the same as N.)

Existence of inverses. The inverse of a coset is found by inverting a representative. That is, $(Na)^{-1} = Na^{-1}$. The proof of this statement is simple: $NaNa^{-1} = Naa^{-1} = Ne$.

$\Box$

Definition 4.3 *If N is a normal subgroup of G, then the set of cosets, with coset multiplication, is called the* quotient group *of G modulo N. It is denoted G/N.*

4.1.4   EXERCISES

1. Assume that you have run your new version of **name_group** from Activity 1. Following is a list of sets, expressed in terms of cosets and the operation **oo**. Express them in terms of elements of G and H and the operation $\circ$. For example,

   H .oo a, a an element of G

   could be expressed as,

```
{h .o a : h in H}
```

(a) (H .oo a) .oo (H .oo b), a,b elements of G

(b) H .oo ab, a,b elements of G

(c) a .oo H .oo i(a), a an element of G

(d) {H .oo a : a in G}.

2. Following is a list of sets expressed in mathematical notation in terms of elements of G and H, where a is an element of G. Write for each set an **ISETL** expression in terms of cosets. For example,

$$\{xa : x \in H\}$$

could be expressed as,

```
H .oo a
```

(a) $\{xay : x, y \in H\}$

(b) $\{xy : x, y \in H\}$

(c) $\{axy : x, y \in H\}$

(d) $\{axya^{-1} : x, y \in H\}$.

3. State and prove the following statements (used in the proof of Proposition 4.1, p. 122) concerning the product of subsets of a group G.

(a) If H is a subgroup of G then $HH = H$.

(b) Associativity: For all subsets R, S, T of G, $(RS)T = R(ST)$.

4. Prove the converse of Proposition 4.1.

5. Prove Proposition 4.2, p. 123 concerning equivalent conditions for normality.

6. Show that the center $Z(G)$ (see p. 94, Exercise 20) of a group G is always normal.

7. What can you say about normality in a commutative group? State and prove an appropriate theorem.

8. Make up and prove other statements equivalent to normality, but not included in Proposition 4.2.

9. Show that the subgroup of rotations is normal in D_n, the group of symmetries of the regular n-gon, but that a subgroup consisting of the identity and one reflection is not.

10. Show that the $n \times n$-matrices with determinant 1 form a normal subgroup in the group of all $n \times n$-matrices with non-zero determinant. (See p. 52).

11. Show that K_2 (see Activity 2, p. 120) is normal in S_4, but K_1 is not. Explain how this can happen with groups that are isomorphic (i.e., the same except for renaming).

12. Prove that a subgroup having only two cosets must be normal.

13. Show that the subgroup A_n of even permutations (p. 105) is normal in S_n.

14. Does A_4 have a normal subgroup of order 3? of order 4?

15. Does S_3 have a normal subgroup of order 3? of order 2?

16. Give an example to show that being a normal subgroup is not a transitive relation; that is, find a chain of subgroups $N \subset H \subset G$ such that N is normal in H and H is normal in G but N is *not* normal in G.

17. Determine all of the normal subgroups of S_3 and S_4.

18. If N and M are normal subgroups of G and $N \cap M = \{e\}$, prove that $nm = mn$ for all $n \in N, m \in M$.

19. Prove that if a finite group has a single subgroup of a certain order, then this subgroup must be normal.

20. Prove that the set of all right cosets of a subgroup N of a group G, with coset multiplication, is a group if and only if N is normal. (We have already proved one direction in Theorem 4.1, p. 126. It is left to prove that if it is a group then N is normal.)

21. What is the order of the element $5 + \langle 6 \rangle$ in the quotient group $\mathcal{Z}_{18}/\langle 6 \rangle$?

22. What is the order of the element $14 + \langle 8 \rangle$ in the quotient group $\mathcal{Z}_{24}/\langle 8 \rangle$?

23. Prove that a quotient group of an Abelian group is Abelian and a quotient group of a cyclic group is cyclic. Give examples to show that the converse is false (that is, a quotient group which is Abelian or cyclic, but the original group is not).

24. Give examples of a subgroup H of G and two elements $a, b \in G$ so that

   (a) $(Ha)(Hb) \neq Hab$ (in the generalized product of subsets of G).

(b) there exist $a', b' \in G$ such that $Ha' = Ha$ and $Hb' = Hb$ but $Ha'b' \neq Hab$.

25. Let G be a finite Abelian group, p a prime number dividing the order of G. Prove that G has a subgroup of order p. (This is a partial converse to Lagrange's theorem, called Cauchy's theorem; it is also true in the non-Abelian case, but the proof there is considerably harder.)

    One approach is to show by induction that G must have an element of order p. If it does not, you can get a contradiction by considering $G/\langle a \rangle$ (where a is any element other than the identity) and using the induction hypothesis.

4.2 Homomorphisms

4.2.1 ACTIVITIES

1. In this section you will be working with maps (functions) from one group to another. You thus need a notation to deal with two groups at the same time. Recall that name_group equips its input group with the standard notation G, o, e, i. Write a new proc name_group' that will equip its input group with the notation G', o', e', i'.

2. The main issue in this section is the study of how applying a map to elements of a group relates to the group operations in the two groups. Specifically, if f is a map from G to G' and a, b are any two elements of G, we can look for a simple relation between their product a .o b (in G) and the product of their images f(a) .o' f(b) (in G'). You will see that sometimes there is and sometimes there is not such a relation.

   In this activity you are to investigate the following list of functions and try to discover this simple relation. To make things more interesting, we have included some examples in which the relationship holds and some in which it does not. Of course you should implement the function in **ISETL** (when that is possible) and run it to help you look at a large number of examples quickly.

   Once you have decided on your relationship, explain carefully why each particular function in the list does or does not satisfy it.

   (a) $f : \mathcal{Z}_{12} \to \mathcal{Z}_{12}$ defined by $f(x) = 3x \bmod 12$.

   (b) $g : \mathcal{Z}_{12} \to \mathcal{Z}_{12}$ defined by $g(x) = x +_{12} 3$.

   (c) $h : \mathcal{Z}_{20} \to \mathcal{Z}_5$ defined by $h(x) = x \bmod 5$.

   (d) $u : \mathcal{Z}_{20} \to \mathcal{Z}_7$ defined by $u(x) = x \bmod 7$.

   (e) $j : S_3 \to S_3$ defined by $j(p) = (1\ 2\ 3)p$.

(f) $k : S_3 \to S_3$ defined by $k(p) = (1\ 2\ 3)p(1\ 3\ 2)$.

(g) $F : S_3 \to S_3$ defined by $(p) = (1\ 2\ 3)p(1\ 2\ 3)$.

(h) $G : S_3 \to S_3$ defined by $G(p) = (1)$ for all $p \in S_3$.

(i) $H : S_3 \to S_3$ defined by $H(p) = (1\ 2)$ for all $p \in S_3$.

(j) $L : S_3 \to \{1, -1\}$ defined by

$$L(p) = \begin{cases} 1 & \text{if } p \text{ is an even permutation} \\ -1 & \text{if } p \text{ is an odd permutation.} \end{cases}$$

(k) $T : S_3 \to S_2$ defined as follows. Given $p \in S_3$, $T(p)$ is the permutation given by,

$$T(p)(i) = \begin{cases} p(i) & \text{if } p(i) \neq 3 \\ 2 & \text{if } p(i) = 3. \end{cases}$$

(l) $U : \mathcal{Z} \to \mathcal{Z}$ defined by $U(n) = 7n$.

(m) $V : \mathcal{Z} \to \mathcal{Z}$ defined by $V(n) = 7n + 2$.

3. Assume that G and G' have been defined by running name_group and name_group' Write an **ISETL** func is_hom that will accept a representation f of a function from G to G'. The func is_hom is to return true or false depending on whether f satisfies the relationship you obtained in the previous activity.

   Check is_hom on the examples of the previous activity.

4. Find two different maps from S_4 into $\{1, -1\}$ for which is_hom will return true.

5. Find two maps from $\mathcal{Z}_{12}$ to $\mathcal{Z}_4$ for which is_hom will return true.

6. Find onto maps from $\mathcal{Z}$ to $\mathcal{Z}_{12}$ and from $\mathcal{Z}$ to $\mathcal{Z}_3$ for which is_hom (run in **VISETL**, of course) will return true.

7. Assume that G and G' have been defined by running name_group and name_group'. Write an **ISETL** func ker that will accept a representation f of a function from G to G'. Your func is to return the set of all elements of G which are mapped by f to the identity e' of G'.

   Run your func on all of the examples in Activities 2, 4, 5, 6.

8. Determine the cases considered in Activity 7 in which the set returned by ker is a subgroup of G.

9. Determine the cases considered in Activity 7 in which the set returned by ker is a normal subgroup of G.

10. In the previous three activities you have computed for a given map the set of all elements which are mapped to the identity. Considering now only those maps in Activities 2, 4, 5, 6 for which is_hom returns true, do the following for each such map $f : G \to G'$.

   (a) Compute, for each element $y \in G'$, the set of all elements $x \in G$ such that $f(x) = y$.

   (b) How many sets do you get?

   (c) For each set, how many elements does it have?

   (d) Do you notice anything familiar happening?

11. Write an **ISETL** func is_iso that has the same structure as is_hom and checks that the map satisfies the same condition as is_hom and, in addition, is both 1-1 and onto.

   Apply your func is_iso to all of the maps in Activities 2, 4, 5, 6.

12. In the previous activity, is there anything to be said about the relationship between the results of ker and is_iso ?

13. Consider the following list of groups. Using reasoning or trial and error with is_iso, find as many pairs G, G' for which there exists a map f such that is_iso(f,G,G') will return true.

   (a) $(\mathcal{Z}_4, +_4)$

   (b) $(\mathcal{Z}_6, +_6)$

   (c) $(\{1, -1\}, *)$

   (d) The Klein 4-group, K.

   (e) S_3

   (f) S_4

   (g) The subgroup of S_4 generated by the cycle (1 2 3 4).

   (h) The subgroup K_1 of S_4.

   (i) The subgroup K_2 of S_4.

   (j) The quotient group S_4/A_4.

14. Suppose we have a function $f : G \to G'$. The inverse image of a subset of G' is the set of all elements of G whose images fall in that subset. The inverse image of a single element $a \in G'$ is taken to mean the inverse image of a. Write a func inv_im to compute inverse images of subsets and of elements of G'. (Assume name_group and name_group' have been run.)

15. Use your new func inv_im to investigate inverse images of various subgroups and single elements of G' in all the examples you worked with in the preceding activities. In particular, there are some very interesting relationships among the inverse images of all the elements of G' for a given example of $f : G \to G'$. See if you can find them. Note that some of this work you may have already done in Activity 10.

16. Make a list of all of the examples of maps $f : G \to G'$ considered in this section for which is_hom returns the value true. Prove or find a counterexample for each of the following statements.

    (a) $f(e) = e'$.

    (b) For all $a \in G$, $f(a^{-1}) = (f(a))^{-1}$.

    (c) If H is a subgroup of G, then the image of H under f, that is, the set $f(H) = \{f(x) : x \in H\}$, is a subgroup of G'.

    (d) If H' is a subgroup of G', then $f^{-1}(H')$ is a subgroup of G, where $f^{-1}(H') = \{x \in G : f(x) \in H'\}$, the inverse image of H' under f.

    (e) Suppose that H is a subgroup of G and $f(H)$ is a subgroup of G'. If H is normal in G, then $f(H)$ is normal in G'.

    (f) Suppose that H is a subgroup of G and $f(H)$ is a subgroup of G'. If $f(H)$ is normal in G', then H is normal in G.

    (g) If G is a cyclic group, then G' is a cyclic group.

    (h) If G' is a cyclic group, then G is a cyclic group.

    (i) Suppose that H is a subgroup of G and $f(H)$ is a subgroup of G'. If H is cyclic, then $f(H)$ is cyclic.

    (j) Suppose that H is a subgroup of G and $f(H)$ is a subgroup of G'. If $f(H)$ is cyclic, then H is cyclic.

    (k) If G is a commutative group, then G' is a commutative group.

    (l) If G' is a commutative group, then G is a commutative group.

    (m) If $a \in G$ has order 4, then so does $f(a)$.

    (n) If $f(a)$ has order 4, then so does a.

    (o) Suppose that both G and G' are permutation groups (that is, subgroups of some S_n). If $a \in G$ is a cycle, then $f(a)$ is a cycle.

    (p) Suppose that both G and G' are permutation groups. If $f(a)$ is a cycle, then a is a cycle.

17. Look at all the statements of Activity 16 for which you found counterexamples. Would any of them become true if you added the assumption that $f : G \to G'$ was "onto" (that is, $f(G) = G'$)? What about one-to-one? Both onto and one-to-one?

4.2.2 Homomorphisms and kernels

A homomorphism is a special kind of map (function) f from one group G to another G'. In Activity 2 you looked for a special relationship between $f(ab)$ and $f(a)f(b)$ for elements a, b of G. Another way of putting this is that if f is to "respect" the group structure of G and G', the relationship between a, b, and ab had better be preserved among their images $f(a)$, $f(b)$ and $f(ab)$. (As usual in mathematical notation, we have dropped the symbol for the operations (o, o') and have used the same notation (concatenation) for the operation in both groups, though it should be remembered that the two operations are, in general, different.)

Hopefully you found that in several examples the image of the product and the product of the images were equal. A map with this property is called a homomorphism. Here is the "official" definition.

Definition 4.4 (homomorphism) *Let G and G' be groups. A function* $f : G \rightarrow G'$ *is called a* homomorphism *if*

$$f(ab) = f(a)f(b) \quad \text{for all } a, b \in G.$$

G' is said to be a homomorphic image *of G if there exists a homomorphism $f : G \rightarrow G'$ which is onto.*

Definition 4.5 (image) *Let $f : G \rightarrow G'$ be a homomorphism. The* image *of an element $a \in G$ is the element $f(a) \in G'$. The* image *of a subset S of G is the subset*

$$f(S) = \{f(a) : a \in S\}$$

of G'. The image of f *is the image of G. The* inverse image *of a subset S' of G' is the subset*

$$f^{-1}(S') = \{a \in G : f(a) \in S'\}$$

Definition 4.6 (kernel) *The* kernel *of a homomorphism $f : G \rightarrow G'$ is the inverse image of the identity in G', that is, the set*

$$\ker(f) = \{a \in G : f(a) = e'\}$$

4.2.3 Examples

In the activities you were asked to find various homomorphisms. Perhaps you discovered that Z_5 is a homomorphic image of Z_{20}, $\{1, -1\}$ is a homomorphic image of both S_3 and S_4, and both Z_3 and Z_{12} are homomorphic images of Z. We can look into some of these examples in more detail.

Homomorphisms from Z to Z_n. In looking for a homomorphism from Z to Z_n (an arbitrary but fixed n), it is helpful to look back at whatever we know about the relationship between these two groups. In Theorem 2.2, p. 63 we established the following:

$$\text{for all } a, b \in \mathcal{Z}, \ \overline{a+b} = \overline{a} +_n \overline{b}$$

where $\overline{x}$ denotes $x \bmod n$. Now, merely translating this relation to the language of maps, shows that the map $f : \mathcal{Z} \to \mathcal{Z}_n$ defined by $f(a) = a \bmod n \ (= \overline{a})$ is a homomorphism. Indeed,

$$f(a+b) = \overline{a+b} = \overline{a} +_n \overline{b} = f(a) +_n f(b).$$

Clearly this homomorphism is onto (every element in $\mathcal{Z}_n$ is the image of some element of $\mathcal{Z}$, e.g. of itself), so that $\mathcal{Z}_n$ is a homomorphic image of $\mathcal{Z}$.

The kernel of this homomorphism is the set of all elements $a \in \mathcal{Z}$ such that $a \bmod n = 0$, that is, the set $n\mathcal{Z}$ of all the multiples of n. Thus, the kernel is a subgroup of $\mathcal{Z}$. It is also normal.

Note that the inverse image of 1 is the set of all integers that leave remainder of 1 upon division by n. Another way of saying this is that the elements with the same image 1 form the coset $\ker(f) + 1$. Compare this to what you observed in doing Activities 10 and 15.

In general, the inverse image of any $a \in \mathcal{Z}_n$ (i.e., the set of all elements of $\mathcal{Z}$ that have a as their image) is the coset $K + a$, where K is the kernel of f. Your work in Activities 10 and 15 was intended to point out several examples of this. We'll return to this important relation in Proposition 4.7.

Homomorphisms from $\mathcal{Z}_n$ to $\mathcal{Z}_m$. The homomorphism from $\mathcal{Z}_{12}$ to $\mathcal{Z}_3$, and in general from $\mathcal{Z}_n$ to $\mathcal{Z}_m$ where $m|n$, is defined similarly to the homomorphisms from $\mathcal{Z}$ to $\mathcal{Z}_n$, that is, for $a \in \mathcal{Z}_n$, $f(a) = a \bmod m$. The proof that this is indeed a homomorphism is also similar, except that you need to use the additional fact that if $m|n$ then

$$(a \bmod n = b \bmod n) \Rightarrow (a \bmod m = b \bmod m).$$

The proof of this last fact, as well that f is a homomorphism, is left as an exercise. Note that in this example, again, the inverse image of an element of $\mathcal{Z}_m$ is a coset of $\ker(f)$ in $\mathcal{Z}_n$.

Homomorphisms from S_n to $\{1, -1\}$. For each n, there are two homomorphisms from S_n to $\{1, -1\}$. One is the "trivial" homomorphism given by $f(p) = 1$ for all $p \in S_n$. (Such a homomorphism, sending the entire group G to the unit element of G', exists between any two groups, hence the denotation "trivial".) The kernel is the whole of S_n. The second, more interesting, homomorphism $f : Sn \to \{1, -1\}$ is defined by the condition that $f(p) = 1$ if p is even and $f(p) = -1$ if p is odd. You can check that this is indeed a homomorphism by using the relevant properties of even and odd permutations (see p. 105). The kernel is A_n, the (normal) subgroup of all even permutations.

4.2.4   Invariants

In Activities 16 and 17 you investigated how various properties in G are or are not preserved when translated to G' via a homomorphism. Since a homomorphism $f : G \to G'$ "respects" the group structure of G and G' (i.e., it preserves the group operation), many properties which are true of elements or subsets of G remain true when these are replaced by their images in G'. Such properties are called *invariants*. On the other hand, there are properties that are not preserved. This happens either because f may fail to be one-to-one (so that different elements in G may have equal images in G'), or it may fail to be *onto*, in which case there are elements of G' about which f says absolutely nothing.

An example of the first kind of "non-invariant" property is the order of an element (the order of $f(a)$ may be smaller than that of a – see Activity 17(m)). To preserve such properties it is necessary that f be one-to-one.

An example of the second kind is commutativity (Activity 17(k)). For instance, we can define a (one-to-one) homomorphism of $f : \mathcal{Z}_4 \to S_4$, mapping the elements of $\mathcal{Z}_4$ to the corresponding powers of the cycle (1 2 3 4). (See the exercises for details.) Now $\mathcal{Z}_4$ is commutative, but the existence of such a homomorphism says nothing about commutativity of the product of elements of S_4 that are not the image of an element of $\mathcal{Z}_4$. To preserve such properties, f is required to be onto.

Each time we conjecture that a certain property is an invariant, we need to prove something, or to find a counterexample. However, here are three "rules of thumb" that will help you get a feeling about invariants and develop the ability to make quick guesses.

1. Properties of elements or subsets of G that can be expressed by *equalities* only will, in general, be true of the images of these elements or subsets in G'. This is because it is okay to apply f to both sides of the equation (can you see why?)

2. Properties that can be expressed by equalities *and quantifiers* only, will in general be preserved provided that f is *onto*. This is because such properties talk about the totality of all the group elements.

3. When the homomorphism f is both one-to-one and onto, then all the "abstract" properties of G are peserved. (This doesn't include properties that refer to the specific form of elements, such as "there exists a matrix with 0 in the upper-left corner".) Such a homomorphism is called isomorphism in Section 4.2.6, and is the formal expression of our earlier ideas about "groups being equal except for renaming" (See Chapter 2, p. 76ff.)

The following proposition in essence states that identities, inverses and subgroups are invariants of any homomorphism. Note that neither the definitions of identity nor subgroup are of the form mentioned in rule 1 above, but have equivalent formulations that are.

Proposition 4.4 *Let $f : G \to G'$ be a homomorphism. Then*

   1. $f(e) = e'$.

   2. *For $a \in G$, $f(a^{-1}) = (f(a))^{-1}$.*

   3. *If H is a subgroup of G then $f(H)$ is a subgroup of G'.*

Proof. We prove the first, leaving the others as exercises. We use the characterization of the identity element in a group as the only element satisfying $x^2 = x$. We start with $e^2 = e$ and apply f to both sides to get $f(e^2) = f(e)$. Using the homomorphism property of f, we have $(f(e))^2 = f(e)$, hence $f(e)$ must be the identity element of G'.

▯

The following proposition lists some invariants of onto homomorphisms. We leave the proof as an exercise, as well as the task of finding counterexamples to show that without the onto assumption these are not invariants.

Proposition 4.5 *Let $f : G \to G'$ be a homomorphism onto G'.*

   1. *If G is Abelian, so is G'.*

   2. *If G is cyclic, so is G'.*

   3. *If H is a normal subgroup of G, then $f(H)$ is a normal subgroup of G'.*

Note that the order of an element is not an invariant even when f is an onto homomorphism. That is because the definition of order (if you look carefully into it) involves an *inequality*. You can show, however, that if f is a one-to-one onto homomorphism (i.e., an isomorphism), then order is preserved.

4.2.5 Homomorphisms and normal subgroups

The next proposition establishes the first important connection between the material of this section and that of the previous section. In the next section you will see that normal subgroups, quotient groups, homomorphisms and kernels are all very closely related.

Proposition 4.6 (kernels are normal.) *The kernel of a homomorphism $f : G \to G'$ is a normal subgroup of G.*

Proof. We give a sketch of the proof and leave the details for the reader. That $\ker(f)$ is closed follows from the property $e'e' = e'$.
The identity element of G is in $\ker(f)$ since $f(e) = e'$.
$\ker(f)$ has inverses since $f(a^{-1}) = (f(a))^{-1}$.

If $h \in \ker(f)$ and $a \in G$, then $aha^{-1} \in \ker(f)$, since

$$
\begin{aligned}
f(aha^{-1}) &= f(a)f(h)f(a^{-1}) = f(a)e'f(a^{-1}) \\
&= f(a)f(a^{-1}) = f(aa^{-1}) = f(e) = e'
\end{aligned}
$$

Thus $\ker(f)$ is normal.

☐

Since kernels of homomorphisms are normal subgroups it seems reasonable to investigate general properties of normal subgroups in the case of a normal subgroup that arises as the kernel of a homomorphism. In Activities 10 and 15 you saw examples in which the inverse image of an element was a coset of the kernel. The following proposition establishes that phenomenon as a general fact for homomorphisms.

Proposition 4.7 (Cosets of kernels) *Let $f : G \to G'$ be a homomorphism of G onto G', K its kernel, and $y \in G'$. Then $f^{-1}(y)$ is a coset of K.*

Proof. We will show that $f^{-1}(y) = Kx$ where x is any element of $f^{-1}(y)$. We will do this in the straightforward manner of checking elements.

If $z \in f^{-1}(y)$, then $f(z) = y = f(x)$. Hence, by the homomorphism property,

$$
f(zx^{-1}) = f(z)f(x^{-1}) = y(f(x))^{-1} = yy^{-1} = e',
$$

so $zx^{-1} \in K$, so $z \in Kx$.

On the other hand, take any $kx \in Kx$. We have

$$
f(kx) = f(k)f(x) = e'y = y,
$$

so $kx \in f^{-1}(y)$.

Thus, the two sets are equal.

☐

An interesting example

This example is motivated by an "inverse problem". It will be taken up more thoroughly in the next section. Instead of defining a homomorphism and then finding its kernel, we look for a homomorphism from S_4 with a given normal subgroup as its kernel. In Activity 2 of Section 4.1, p. 120, you saw that the group K_2 given by

$$
K_2 = \{I, (1\ 2)(3\ 4), (1\ 3)(2\ 4), (1\ 4)(2\ 3)\}
$$

is a normal subgroup of S_4. By Proposition 4.6 it is *possible* for K_2 to be the kernel of some homomorphism of S_4. Can we actually find such a homomorphism?

According to Proposition 4.7, the cosets of the kernel are the inverse images of single points. Since, in our case, S_4 has 24 elements and K_2 has 4 elements, we can conclude (see Section 3.3.2) that the image will have 6 elements. This heuristic reasoning leads us to look for a homomorphism from S_4 onto S_3 (which has order 6). (Why not Z_6?)

To do this we must find a reasonable way of transforming a permutation p on the set $\{1, 2, 3, 4\}$ to a permutation $f(p)$ on the set $\{1, 2, 3\}$. If $p(4) = 4$, then we have no problem because then p maps the set $\{1, 2, 3\}$ into itself so we can just restrict p to $\{1, 2, 3\}$. If $p(4)$ is anything else, we can first convert it to a permutation which maps 4 to itself by multiplying by one of the elements of K_2.

Let's look at that last point a little more closely. We can write

$$K_2 = \{k_1, k_2, k_3, k_4 = I\}$$

where k_j contains the transposition $(j\ 4)$. It then follows that if $p(4) = j$, the permutation $k_j p$ maps 4 to itself.

Thus we define our map $h : S_4 \to S_3$ as follows. If $p \in S_4$, then $h(p)$ is the restriction of $k_j p$ to $\{1, 2, 3\}$ where $j = p(4)$. From what we have just said, $h(p)$ is an element of S_3. It is necessary to prove that it is a homomorphism and that it is onto S_3. We leave this for the exercises.

The interesting part is to determine the kernel of h. Just exactly when will a permutation $p \in S_4$ be mapped by h to the identity permutation in S_3? That is, when will $k_j p$ $(j = p(4))$ be the identity on $\{1, 2, 3\}$? It is necessary to look at the permutation k_j. This permutation consists of two transpositions. One interchanges j and 4 and the other interchanges the other two elements of $\{1, 2, 3, 4\}$. This means that in addition to $p(4) = j$, if $h(p)$ is to be the identity, it must be that p interchanges the other two elements of $\{1, 2, 3, 4\}$, that is, the two that are not j or 4. Thus we have that p maps j to 4 and interchanges the other two elements. Since it is a permutation, it must map 4 to j, that is p interchanges j and 4 as well as the other two elements. This is precisely an element of K_2, which consists of all permutations in S_4 that are the product of two disjoint transpositions.

Remark. In general, the inverse problem to finding the kernel of a given homomorphism is the following: Given a group G and a normal subgroup N, find a homomorphism with N as its kernel; that is, find a group G' and a map $f : G \to G'$ so that f is a homomorphism and $\ker(f) = N$. In the next section we will give a general solution to this problem.

4.2.6   Isomorphisms

A special case of homomorphism is an isomorphism, which in general allows us to view two different groups as identical except for the names of their elements and of the group operation. The definition of isomorphism is motivated by the desire that it preserve all abstract group properties, i.e., those properties that are not concerned with the particular name or form of the elements.

Definition 4.7 (isomorphism.) *A one-to-one onto homomorphism* $f : G \to G'$ *is called an* isomorphism. *A group G is said to be* isomorphic *to a group G' if there exists an isomorphism from G to G'.*

In Activities 17 and 14 you considered various group properties and the question of whether they were preserved by homomorphisms and isomorphisms. Presumably you found that most, but not all of the properties were preserved by homomorphisms and even more were preserved by isomorphisms.

Remark. The relation "is isomorphic to" is reflexive, symmetric and transitive. That is, every group is isomorphic to itself (via the identity map); if G is isomorphic to G' then G' is isomorphic to G (via the inverse map); and if G is isomorphic to G' and G' is isomorphic to G'', then G is isomorphic to G'' (via the composition of the two maps). As a result we can talk about two groups being isomorphic to each other. Also, two groups which are isomorphic to a third group are isomorphic to each other. (Compare Exercise 8, p. 55 in Proposition 3.6, p. 90 and Exercise 11, p. 116.)

4.2.7 Identifications

This section (in fact, much of group theory) is about processes of identification; that is, how can we take two things that are different and look at them as if they were the same? (This is what we mean by "identifying" them or making them identical.) We can do this in at least two ways.

The basic idea is that a map (function) from one group to another establishes a certain connection between the two groups. The connection may be more or less intimate (preserving more or less of the properties of the original group) depending on the requirements we make on the map. It turns out that everything we want can be achieved by making one or another of three requirements — one-to-one, onto, and preserving the group operation.

For example, we may decide to identify all the integers which leave the same remainder when divided by 12, ignoring for this purpose all their other differences. (A more technical way of saying this is that we want to turn congruence mod 12 into equality.) As usual, the process is in our mind; the mathematical expression of this process is in a function having certain properties. Indeed, in Activity 5 you found a homomorphism $h : \mathcal{Z} \to \mathcal{Z}_{12}$, presumably the one given by $h(n) = n \bmod 12$. The effect of this homomorphism is to "identify" those elements of $\mathcal{Z}$ which have the same remainder on division by 12. The identification takes the form of mapping all such elements to the same element in $\mathcal{Z}_{12}$. You can reverse the process of identification by looking at the inverse image of elements of $\mathcal{Z}_{12}$. The fact that this is always a coset of the kernel gives this particular process of identification an algebraic structure that can be usefully applied to other questions in group theory. In fact, the algebraic structure accounts for the fact that the set of elements identified with the unit element (i.e the kernel), entirely determines all other identifications (i.e., the cosets of the kernel).

In general, the effect of a homomorphism $h : G \to G'$ is to identify certain elements of G by mapping them to the same element of G'. The set of all things identified with a particular $x \in G$ is the coset Kx of x with respect to the normal subgroup K of all those things "identified" with the identity in G (mapped to the identity in G'). Moving from G to G' is a way of deciding (for a particular investigation) that elements which are similar in some sense will be treated as equal. Again, the reverse of this process is captured by the cosets of the kernel, K.

Another way of identifying things is the case of two different groups being considered "the same except for renaming", in which we identify each element of one group with a corresponding element of the other. This is captured by the concept of isomorphism which is a map that preserves everything related to group properties of any kind. An interesting consequence of the mathematics that does this is that the idea of preserving "everything related to group properties" is implemented by only three requirements. The map must be one-to-one, it must be onto, and it must preserve the operation, that is,

$$f(ab) = f(a)f(b).$$

Remark. One-to-one and kernels. There is a simple test for a homomorphism being one-to-one: the condition that the kernel be reduced to the identity (the smallest possible set since the identity will always be an element of the kernel). We leave the proof of this fact for the exercises, but it is interesting to observe that this condition can be interpreted as saying that there is no identification of the first kind. In other words, if we think of two elements being identified if they are mapped to the same element, then a map which is one-to-one will only identify an element with itself. In this sense, the two kinds of identification are mutually exclusive.

You investigated several examples of isomorphisms in the activities. For instance, in Activity 16 you were looking for isomorphisms between pairs of groups. Did you see that the two groups $\{1, -1\}$ and S_4/A_4 are isomorphic? Surely, two objects could not be more different than the group of two numbers 1,-1 (with the operation of ordinary multiplication) and the quotient group of the set of all cosets of A_4 in S_4 (with the operation of coset multiplication). Nevertheless, the fact that there is an isomorphism between them says that as groups, they are the same except for renaming. This means that you will never see a "group phenomenon" occur in one and not the other.

4.2.8 EXERCISES

1. Prove the rest of Proposition 4.4, p. 136: Let $f : G \rightarrow G'$ be a homomorphism. Then

   (a) For $a \in G$, $f(a^{-1}) = (f(a))^{-1}$.

   (b) If H is a subgroup of G then $f(H)$ is a subgroup of G'.

2. In the previous exercise, if we assume in addition that H is normal in G, does it follow that $f(H)$ is normal in G'? In $f(G)$?

3. Prove Proposition 4.5, p. 136 on invariants of onto homomorphisms. Show (by example) in each case that the "onto" assumption is indeed necessary.

4. Let G, G' be any two groups and define a map $t : G \rightarrow G'$ by $t(g) = e'$ for all $g \in G$. Show that t is a homomorphism. What is its kernel? (t is called a *trivial* homomorphism.)

5. Complete the proof that the map h defined beginning on p. 137 is a homomorphism of S_4 onto S_3.

6. Prove or find counterexamples for everything in Activities 13 and 16, p. 131.

7. Prove that all the "naive isomorphisms" we encountered in this book are indeed isomorphisms according to the formal definition given here.

8. Consider the map $x \mapsto \cos x + i \sin x$ (also denoted e^{ix}), $0 \leq x < 2\pi$. Show that this map is an embedding (i.e. one-to-one homomorphism) of the additive group $[[0, 2\pi), +_\pi]$ in the multiplicative group $[C - \{0\}, *]$. What is its image? Can you think of some geometric meaning for this map?

9. Show that the map sending each complex number to its modulus (or absolute value) is a homomorphism of $[C - \{0\}, *]$ onto the multiplicative group of positive real numbers. What is its kernel?

10. Show that the map $A \mapsto \det A$ is a homomorphism of the group of real $n \times n$-matrices with non-zero determinant onto the multiplicative group of non-zero real numbers. What is its kernel?

11. Prove that the additive groups Z and Q are not isomorphic.

12. Find non-trivial homomorphisms $Z_3 \rightarrow Z_{15}$ and $Z_{15} \rightarrow Z_3$.

13. Find non-trivial homomorphisms $S_3 \rightarrow Z_{12}$ and $Z_{12} \rightarrow S_3$.

14. Let G be any group, $x \in G$ any element. Show that there is a unique homomorphism of $\mathcal{Z}$ into G which maps 1 to x. Find a condition on x for this homomorphism to be onto, and a condition for it to be one-to-one.

15. How many homomorphisms are there:

   (a) of $\mathcal{Z}$ onto $\mathcal{Z}$? of $\mathcal{Z}$ into $\mathcal{Z}_2$? of $\mathcal{Z}$ onto $\mathcal{Z}_2$?

   (b) of $\mathcal{Z}$ into $\mathcal{Z}_8$? of $\mathcal{Z}$ onto $\mathcal{Z}_8$?

   (c) of $\mathcal{Z}_{12}$ onto $\mathcal{Z}_5$? of $\mathcal{Z}_{12}$ into $\mathcal{Z}_6$? of $\mathcal{Z}_{12}$ onto $\mathcal{Z}_6$? of $\mathcal{Z}_{12}$ into $\mathcal{Z}_{16}$?

16. Let I be any interval on the real line and let G_n be the point-wise additive group of all real-valued functions which are n times differentiable on I. Show that the derivative map $f \mapsto f'$ is a homomorphism of G_n to G_{n-1}. What is its kernel? Is it onto?

17. Give a full definition of the map $f : \mathcal{Z}_4 \rightarrow S_4$ described on p. 135, and prove that it is a homomorphism. Conclude that commutativity is not an invariant under homomorphisms.

18. Prove that a homomorphism f is one-to-one if and only if $\ker f = \{e\}$.

19. Prove that the relation "is isomorphic to" among groups is an equivalence relation. (See the remark following Definition 4.7, p. 139.)

20. An isomorphism of a group G onto itself is called an *automorphism of G*. Prove that the set of all automorphisms of G forms a group relative to the operation of composition of functions. We shall denote this group as $A(G)$.

21. Show that the map $x+yi \mapsto x-yi$ (this is called *complex conjugation*) is an automorphism of $[\mathcal{C}, +]$ and also of $[\mathcal{C} - \{0\}, *]$.

22. Let G be a group, $a \in G$, and define a map $f_a : G \rightarrow G$ by $g \mapsto aga^{-1}$ (that is, f_a is conjugation by a). Show that f_a is an automorphism of G. (f_a is called the *inner automorphism* induced by a.)

23. Prove that the subset $I(A) = \{f_a : a \in G\}$ is a subgroup of $A(G)$. Does this subgroup have any connection to G itself?

24. Prove that the map $\phi : a \mapsto f_a$ is a homomorphism of G onto $I(G)$, and determine its kernel.

25. A *permutation* of a set A is a one-to-one map of A onto itself. Prove that the set S_A of all the permutations of A forms a group under the operation of composition of functions.

26. In the context of the previous exercise, show that if A is a finite set with n elements, then S_A is isomorphic to S_n.

27. Let G be a group, $a \in G$, and define a map $l_a : G \to G$ by $g \mapsto ag$. Prove that l_a is a permutation of G.

28. Prove that the map $\psi : a \mapsto f_a$ is an embedding (i.e., one-to-one homomorphism) of G into $A(G)$.

29. Prove Cayley's Theorem: Every finite group is isomorphic to a subgroup of S_n for some n.

4.3   The homomorphism theorem

4.3.1   ACTIVITIES

1. **Review activities.** Let G = Z12, H = {0,4,8}. Run the following statements in your mind (and write down the result) before running them on the computer.

   (a) G := Z12; H := {0,4,8};

   (b) a := 5; H .oo a subset GmodH;

   (c) (H .oo a) subset G;

   (d) (H .oo a) in G;

   (e) (H .oo a) in GmodH;

   (f) H subset G;

   (g) GmodH subset G;

   (h) b := 7;

   (i) K(a); K(b); K(a .o b);

   (j) K(a .o b) = K(a) .oo K(b);

   (k) is_normal(H); (This has not been defined previously, so define it.)

   (l) is_group(GmodH);

   (m) #G, #H, #GmodH, #(H .oo a);

2. In the context of Activity 1, is GmodH a homomorphic image of G?

3. In the context of Activity 1, is GmodH isomorphic to Z4? To the subgroup {0,3,6,9} of Z4?

4. Following is a list of pairs G, N where G is a group and N is a normal subgroup. For each pair, determine whether GmodN is a homomorphic image of G (that is, whether there exists a homomorphism of G onto GmodN). In the case that it is, construct such a homomorphism. Where possible, use your func is_hom (p. 130) to check this.

(a) Z12, {0,4,8}

(b) Z20, {0,4,8,12}

(c) Z, the set of all integer multiples of 17

(d) S3, A3

(e) S4, A4

(f) S4, K2

5. Find the kernels of all the homomorphisms you constructed in the previous activity.

6. Recall the homomorphism f from Z20 to Z5 of Section 4.2.1, Activity 2(c), p. 129 and the func inv_im ("inverse image") Section 4.2.1 Activity 14(c), p. 131. Construct in **ISETL** the normal subgroup N = ker(f), and the quotient group GmodN. Write the multiplication tables of Z5 and of GmodN where G is Z20. Compare the two tables.

7. In the context of the previous activity, apply inv_im to the elements of Z5 and then on those of Z20modN (using the homomorphism of Activity 3(b)) and compare the results. Did you find anything noteworthy?

8. Repeat Activity 7 for the homomorphisms from S4 onto {1,-1} and from S4 onto S3.

9. Repeat Activity 7 for the homomorphism f from Z12 to Z12 by f(x) = 3x mod 12 in Section 4.2 Activity 2(a), p. 129. How is this different from the previous two activities?

10. Using the same homomorphism f and its kernel N from Activity 4(b), find a,b in Z20 such that f(a) = f(b). Find a,b in Z20 such that Na = Nb. Can you find a,b in Z20 such that the first equality holds but not the second? Can you find a,b in Z20 such that the second equality holds but not the first?

11. Determine general conditions on a homomorphism f from G to G' guaranteeing the truth of the following statement for a group G and a normal subgroup N.

    forall a,b in G | f(a) = f(b) if and only if N .oo a = N .oo b

12. Let h be the homomorphism from S4 onto S3 with kernel K2 described in Section 4.2. Implement this homomorphism in **ISETL** and perform the following statements.

```
L := arb(S4/K2);
a:= arb(L);
b := arb(L);
h(a) = h(b);
```

Repeat the same sequence of assignments several times over. Did you notice anything? Can you explain?

13. Find as many of the homomorphic images of the group S3 as you can.

4.3.2 The canonical homomorphism

The focus of this chapter has been the relations between three major constructs of group theory: normal subgroups, homomorphisms and quotient groups. The main relations we have established so far are:

- The quotient group G/N is defined if and only if N is normal (Section 4.1).

- The kernels of homomorphisms are normal subgroups (Section 4.2).

The second of these relations can be interpreted to mean that every homomorphism gives rise to a normal subgroup, namely its kernel. In this section we will show that the converse is also true, that is, every normal subgroup N gives rise to a homomorphism with N as its kernel. Also, we'll tie together all three constructs by formulating theorems connecting them.

In Activity 2 you investigated cases in which G/N is a homomorphic image of G. Hopefully you discovered that this is always the case. In fact, the required homomorphism was there all along in the **ISETL** activities, but we didn't explicitly refer to it as such. We are talking about the func K which assigns to every a in G its coset N .oo a. Now clearly K represents a function from G onto G/N (represented by GmodN). Furthermore, the formula for the product in G/N is

$$(Na)(Nb) = N(ab)$$

which, when formulated in terms of the function K, reads:

$$K(a)K(b) = K(ab).$$

Thus K is a homomorphism. Hopefully you also found what the kernel of K was in several examples when you did Activity 3. We summarize all this in the following theorem.

Theorem 4.2 (The canonical homomorphism.) *Let G be a group, N a normal subgroup of G. Then the map $K : G \to G/N$ defined by $K(a) = Na$, for all $a \in G$, is a homomorphism of G onto G/N, whose kernel is N.*

Proof. The first part of the proof is essentially a repetition of the discussion preceding the theorem, formulated in more precise terms, and using standard mathematical notation. The map K is defined in the statement of the theorem and we need to check that it is a homomorphism. According to Proposition 4.1 we have,

$$K(ab) = Nab = (Na)(Nb) = K(a)K(b),$$

so K is a homomorphism.

Next we observe that K is onto. In fact, if L is any coset and a is any element of L then $L = Na$, hence $L = K(a)$.

Finally, we show that the kernel of K is N. We recall two facts about quotient groups. First, N $(= Ne)$ is the unit element of G/N (see the proof of Theorem 4.1, p. 126); and second, $Na = N$ if and only if $a \in N$ (Lemma 3.1, p. 114.)

Indeed, $\ker(f)$ is the set of all the elements a in G such that $K(a) = N$, that is, all a such that $Na = N$, that is, precisely N.

$$\square$$

Note carefully the double role that N played in the above proof. On the one hand it was an element of G/N, indeed, the identity element there. At the same time, N was a subset of G, in fact, precisely the subset we showed to be the kernel of K.

Definition 4.8 (canonical homomorphism.) *The homomorphism assigning to each $a \in G$ the coset Na in G/N is called the* canonical homomorphism *of G onto G/N.*

Remark. Theorem 4.2 provides a general solution of a problem that was quite difficult in Section 4.2. Recall (p. 137) that we wanted to find a homomorphism of S_4 whose kernel was the normal subgroup K_2. Theorem 4.2 gives it immediately as the canonical homomorphism $K : S_4 \to S_4/K_2$. We can also obtain the same information as in Section 4.2 by checking that S_4/K_2 is isomorphic to S_3. You will have a chance to work out the details in the exercises. It will also follow from Theorem 4.3 below.

Corollary 4.1 *Let S be a subset of a group G. Then S is a normal subgroup of G if and only if S is the kernel of some homomorphism $f : G \to G'$.*

Proof. We saw in the previous section that all kernels are normal subgroups, and Theorem 4.2 establishes that each normal subgroup is a kernel of the corresponding canonical homomorphism.

$$\square$$

Thus in a way we have shown that normal subgroups and homomorphisms are equivalent. We shall next establish a similar relation between homomorphisms and quotient groups; that is, having shown that every quotient group G/N is a homomorphic image of G, we shall now show that, conversely, every homomorphic image of G is (isomorphic to) a quotient group of G.

4.3.3   The fundamental homomorphism theorem

It might be helpful to recall at this stage the activities in which you started with a homomorphic image G' of G (for example, $\mathcal{Z}_5$ of $\mathcal{Z}_{20}$ in Activity 4), then constructed the kernel N of the homomorphism (in this case it is the set of multiples of 5) and the quotient group G/N; then you studied the relation between G/N and G'. Hopefully you noticed that they were "the same except for renaming", that is, isomorphic.

Another point to remember is that the image of a homomorphism $h :$ $G \to G'$, which we shall write as $\text{Im}(h)$ is a subgroup of G' (Proposition 4.4).

Theorem 4.3 (The fundamental homomorphism theorem.) *Let G and G' be groups, h a homomorphism of G onto G', $N = ker(h)$. Then G/N is isomorphic to G'.*

Proof.

> The basic idea of the proof is that, since both G' and G/N are homomorphic images of G, elements in both groups have "sources" in G. We can thus try to define the isomorphism by relating elements with the same source. The difficulty is that each element (both in G' and in G/N) may have many sources, so that this process may not define a function (because the values assigned by it may not be uniquely determined). The crucial fact that comes to our aid is that not only are G/N and G' both homomorphic images of the same group G, but these homomorphisms also *have the same kernel*. In the proof we will use this fact to show that we have actually defined a function and that it is one-to-one. The rest of the proof then consists of rigorous but quite routine checking that the various claims of the theorem indeed follow from the many definitions involved.

We define a function h' from G/N to G' (h' is our candidate for the isomorphism) as follows. Given an element L of G/N, i.e., a coset of N in G, we start by choosing any representative $a \in L$. (Recall that this means that $L = Na$.) We now define $h'(L) = h(a)$ — an element in G'. (See Figure 4.1.)

The definition of h' can be written briefly as $h'(a) = h(a)$. We can thus see that the map h' is induced by the map h. We first show that this actually defines a function.

What we are about to do is a general version of the situation you observed in Activity 10. The issue is that different choices of $a \in L$ could lead to different values of $h'(L)$. However, we shall now show that $Na = Nb$ if and only if $h(a) = h(b)$, from which it follows that different representatives of the same coset L can only lead to different representations of the same element $h'(L) \in G'$, so that h' as defined above is indeed a function. (Compare this with what you saw in Activities 7 and 8.)

> Referring to our opening discussion, this shows that, although each element in G' or G/N may have many "sources" in G, the "degree of freedom" in choosing a source is the same for the two groups. As hinted in the discussion, this not only guarantees that h' is defined but will also establish that h' is one-to-one.

Indeed,

$$
\begin{aligned}
Na = Nb \iff & ab^{-1} \in N \\
\iff & h(ab^{-1}) = e' \text{ (since N is the kernel of h)} \\
\iff & h(a)h(b)^{-1} = e' \\
\iff & h(a) = h(b).
\end{aligned}
$$

Now suppose that $h'(L) = h'(M)$ for some cosets L and M, and let $L = Na$, $M = Nb$. Then $h(a) = h(b)$; hence, $Na = Nb$, i.e. $L = M$. This means that h' is one-to-one. It is also onto since h is onto (can you see the connection?). Finally, h' is a homomorphism since (once again assuming K,M are any cosets and $K = Na$, $M = Nb$),

$$h'(KM) = h'((Na)(Nb)) = h'(N(ab)) = h(ab) = h(a)h(b) = h'(K)h'(M)$$

(The last-but-one equality utilizes the fact that h is a homomorphism.)

We have proved that h' is a one-to-one onto homomorphism, that is, an isomorphism.

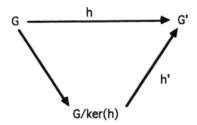

FIGURE 4.1. Isomorphism of $G/\ker(h)$ and G'

The following corollary is useful when we want to determine what a certain quotient group looks like. A standard way to make such a determination is by identifying the given quotient group with some familiar group.

Corollary 4.2 *To show that a quotient group G/N is isomorphic to some group G', it is enough to find a homomorphism of G onto G' with kernel N.*

Corollary 4.3 *To find (up to isomorphism) all groups that are homomorphic images of a group G, it is enough to find all the normal subgroups N of G and form all the quotient groups G/N.*

Remark. This corollary significantly simplifies the task of determining all homomorphic images of G. (You had a chance to see, in Activity 11, how difficult this can be without the aid of any theory.) Instead of searching among all the groups in the world, we need only look for normal subgroups N inside G. Then, all homomorphisms of G to a group G' are given by compositions $f \circ h$ where $h : G \to G/N$ is the canonical homomorphism of G onto the quotient group G/N and f is an isomorphism of G/N into a group G'.

On the other hand, even the reduced task of determining all normal subgroups of a given group may prove rather slippery.

Example 1: The integers. We first use Corollary 4.2 to determine all the quotient groups of $\mathcal{Z}$, then Corollary 4.3 to determine all its homomorphic images. We have seen in Theorem 3.3, p. 100, that all the subgroups of $\mathcal{Z}$ are the sets $n\mathcal{Z}$ for $n = 0, 1, 2, \dots$ Since $\mathcal{Z}$ is commutative, these are also all the normal subgroups. We now show that $\mathcal{Z}/n\mathcal{Z}$ is isomorphic to $\mathcal{Z}_n$. Indeed, we already know that the function $a \to a \bmod n$ is a homomorphism of $\mathcal{Z}$ onto $\mathcal{Z}_n$ with kernel $n\mathcal{Z}$ (p. 133); hence, it follows from the Fundamental Theorem of Homomorphisms (via Corollary 4.2) that $\mathcal{Z}/n\mathcal{Z}$ is isomorphic to $\mathcal{Z}_n$.

It now follows that the groups $\mathcal{Z}_n$ are all the homomorphic images of $\mathcal{Z}$, since by Corollary 4.3 every homomorphic image is isomorphic to a quotient group, hence to some $\mathcal{Z}_n$. (We should add that $\mathcal{Z}$ itself is also a homomorphic image of itself, isomorphic to the quotient group $\mathcal{Z}/\{0\}$.)

Example 2: The modular groups. Again we use Corollaries 4.2 and 4.3 to determine the quotient groups and then all homomorphic images of $\mathcal{Z}_n$. As with $\mathcal{Z}$, Theorem 3.3 tells us that the (normal) subgroups of $\mathcal{Z}_n$ are the sets $k\mathcal{Z}_n$ where $n = km$, and the remark on p. 134 implies that the corresponding homomorphism is the function $a \to a \bmod k$ which is onto $\mathcal{Z}_k$. Thus, by the Fundamental Theorem, $\mathcal{Z}_n/k\mathcal{Z}_n$ is isomorphic to $\mathcal{Z}_k$.

Therefore we conclude that the groups $\mathcal{Z}_k$, $k|n$, are all the homomorphic images of $\mathcal{Z}_n$.

Example 3: The symmetric groups S_n. It can be shown that for $n \neq 4$ the only non-trivial normal subgroup of S_n is A_n, the group of all even permutations. (This is a somewhat involved argument which would take us beyond the scope of this book. See J. Gallian, *Contemporary Abstract Algebra*,

p. 350 for the special case of A_5.) S_4 is an exception, since it has one additional normal subgroup, namely $K_2 = \{I, (1\ 2)(3\ 4), (1\ 3)(2\ 4), (1\ 4)(2\ 3)\}$. Thus, the only non-trivial homomorphic images of S_n for $n \neq 4$ are S_n/A_n, which, in every case, is isomorphic to the group $\{1, -1\}$ with ordinary multiplication. (Prove this with Corollary 4.2, using the homomorphism from S_n to $\{1, -1\}$ from the activities in Section 4.2.) As for S_4, it has one more homomorphic image, namely S_4/K_2, which is isomorphic to S_3 (p. 137).

Hence, the solution to Activity 11 is that the homomorphic images of S_4 are, first of all, as always, S_4 itself and the trivial group $\{I\}$ and then the two groups $\{1, -1\}$ (a homomorphic image of every S_n), and S_3 (a unique case).

4.3.4   EXERCISES

1. Show directly that S_3 is a homomorphic image of S_4 with kernel K_2 by showing that S_4/K_2 is isomorphic to S_3 and using the canonical homomorphism.

2. Let N be a normal subgroup of a finite group G. Prove that the order of Ng in the quotient group G/N divides the order of g in G.

3. Show that Q/Z is an infinite group in which every element has finite order. Show that Z_n can be embedded in Q/Z for all positive integers n.

4. Let G be an Abelian group, n a positive integer. Define two subsets of G as follows: $G_n = \{g \in G : g^n = e\}$ and $G^n = \{g^n : g \in G\}$. Prove that G_n is a normal subgroup of G, that G^n is a group, and that G/G_n is isomorphic to G^n.

5. Assume k, n are integers and k divides n. Prove that Z_n/kZ_n is isomorphic to Z_k.

6. Assume k, n are integers and k divides n. Find a familiar group which is isomorphic to $(Z/nZ)/(kZ/nZ)$. This is a special case of a general theorem about a chain of normal subgroups $N \subset K \subset G$. Formulate and prove the theorem.

7. What can you say about the order of a finite group G if it is known that both Z_{10} and Z_{15} are homomorphic images of G?

8. What can you say about the order of a group G if it is known that for all primes p, Z_p is a homomorphic image of G?

9. Prove that if N and N' are normal subgroups of G and $N \subseteq N'$, then G/N' is a homomorphic image of G/N. What is the kernel of this homomorphism?

10. Let G be the set of all real-valued functions defined on some interval I relative to point-wise addition of functions, and let N be the subset of all functions that have the value 0 on some fixed point $x_0 \in I$. Show that N is a normal subgroup of G and find a familiar group isomorphic to G/N.

11. Let G be the multiplicative group of non-zero real numbers, $N = \{1, -1\}$. Find a familiar group isomorphic to G/N.

5
Rings

5.1 Rings

5.1.1 ACTIVITIES

1. In arithmetic, there is the distributive law involving addition and multiplication: for all numbers a, b, c,

$$a(b + c) = ab + ac$$

$$(b + c)a = ba + ca.$$

Write an **ISETL** func is_dist that accepts a set and two binary operations and returns the truth or falsity of this relation with addition and multiplication replaced by the two operations.

For each of $\mathcal{Z}_n$, $n = 2, 5, 6, 12$, decide on a pair of binary operations and run is_dist on them.

2. Set up a pair of binary operations on each of the sets, $\mathcal{Z}$ (integers), $2\mathcal{Z}$ (even integers), $\mathcal{Z} - 2\mathcal{Z}$ (odd integers), $\mathcal{R}$ (real numbers), $\mathcal{Q}$ (rational numbers) and $\mathcal{C}$ (complex numbers) and, using **VISETL**, run is_dist in each case.

3. Do the same as in Activity 2 for the set S_3.

4. Write an **ISETL** func are_zerodiv that will accept a set, two binary operations on that set for which is_dist returns true, and two

elements of the set. It is assumed that the first operation has an identity which we shall call 0. Your func is to return the truth or falsity of the statement that the two elements are non-zero, but the result of applying the second operation is equal to 0.

Use your are_zerodiv to write a func has_zerodiv which accepts a set and a pair of binary operations, in which the first is assumed to have an identity. The func decides whether or not the set has two elements for which are_zerodiv will return true.

Test has_zerodiv on all of the examples in Activity 1.

5. Write a func units which will accept a set and two binary operations on the set. Your func assumes that the second operation has an identity. It returns the subset of all elements of the set which have an inverse with respect to the second operation.

   Apply units to all of the examples in Activity 1. Can you find an example in which the result returned by units is the empty set? A set with a single element? All but one element of the original set? All of the original set?

6. Write an **ISETL** func is_ring that accepts a set and two binary operations and returns true provided that the set together with the first operation (called addition) forms a commutative group, the set together with the second operation (called multiplication) satisfies closure and associativity, and is_dist returns true.

   Test is_ring on all of the examples in Activity 1.

7. Write an **ISETL** func is_integral_domain that accepts a set and two binary operations and returns true provided that is_ring returns true, the second operation (multiplication) is commutative and has an identity, and has_zerodiv returns false.

   Test is_integral_domain on all of the examples in Activity 1. List all examples in which one of is_ring and is_integral_domain returns true while the other returns false.

8. Write an **ISETL** func is_field that accepts a set and two binary operations (called addition and multiplication) and returns true provided is_ring returns true, the second operation is commutative and has an identity, and every element other than 0 (the identity element with respect to addition) has an inverse with respect to multiplication.

   Can you write an equivalent func not using is_ring, but using is_group *twice* instead?

   Test is_field on all of the examples in Activity 1. List all examples in which one of is_field and is_integral_domain returns true while the other returns false. Do you notice anything remarkable?

9. Edit your **proc name_group** to obtain a **proc name_ring** that will result in the following notation being defined: R, ad, mul, z, neg.

10. Write an **ISETL proc matring** which will assume that **name_ring** has been run. Your **func** will construct from R a new set MR and from ad and mul, two binary operations, Mad and Mmul. The set will consist of all **tuples** of length 2, each of whose elements is a **tuple** of length 2 whose components are elements of R. These elements are to be considered as 2×2 matrices of elements of R, the first **tuple** corresponding to the first row and the second **tuple** corresponding to the second row. The ring addition will be term-by-term addition of matrices and the multiplication will be multiplication of matrices.

Apply **matring** to all of the examples in Activity 1, and then apply is_ring, is_integral_domain, and is_field to the result. What can you say about "preservation"? That is, to what extent is it the case that if one of these **funcs** returns **true** for the original system, then it and/or the others will return **true** for the result of **matring**.

You will probably find in this activity that many of the operations take a long time in **ISETL** so you might have to use **VISETL** for some of them. Probably the most efficient thing to do will be to form a partnership between **ISETL** and **VISETL**.

11. Following is a list of "facts" that are true for ordinary arithmetic. Look at an arbitrary system of a set and two binary operations for which is_ring returns **true** and decide if the statement is true or false. You may use **ISETL**, **VISETL**, or your mind.

In the following, the first operation is denoted by + and the second by concatenation. We denote by 0 and 1 the identities of the first and second operations respectively. We denote by $-$ the inverse operation for the first operation and we use $a - b$ as shorthand for $a + (-b)$.

(a) For all a in the set, $a0 = 0a = 0$.

(b) For all a, b in the set, $a(-b) = (-a)b = -(ab)$.

(c) For all a, b in the set, $(-a)(-b) = ab$.

(d) For all a, b, c in the set, $a(b-c) = ab - ac$ and $(b-c)a = ba - ca$.

(e) For all a in the set, $(-1)a = a(-1) = -a$.

(f) For all a, b in the set, if $ab = 0$ then either $a = 0$ or $b = 0$.

(g) $0 \neq 1$.

(h) For all a, b, c in the set, if $a \neq 0$ and $ab = ac$ then $b = c$.

(i) For all a, b in the set, $(a + b)^2 = a^2 + 2ab + b^2$.

(j) For all a, b in the set, $(ab)^2 = a^2 b^2$.

Look over all the results you obtained in this activity and try to formulate some general relationships. In particular, when a statement is false, try to think of modifications that will make it true.

5.1.2 Definition of a ring

In mathematics, when two entities (such as binary operations) are considered simultaneously, there is generally some property that makes a connection between them. Often this is referred to as a *compatibility condition*. In the case of binary operations, the most common compatibility condition is the distributive law.

Suppose that S is a set with two binary operations. Let us call the first *addition*, and the second *multiplication*. Generally, addition will be indicated by the usual symbol, $+$. For multiplication the symbol will vary with the example and often it will be omitted. As in arithmetic, multiplication can be indicated by juxtaposition, that is, placing the two elements next to each other as in ab. This is often done in a statement about a system in general. When the operation has a name, it is generally used. Thus we would write $a + b$ and ab respectively for addition and multiplication in general, but $+_n$ and $\cdot_n$ for addition and multiplication in $\mathcal{Z}_n$.

Sometimes, a system consisting of a set and two binary operations is written using a triple as in $[S, +, \circ]$ or $[\mathcal{Z}_3, +_3, \cdot_3]$.

Since the general multiplication we will be considering may not be commutative, we will need *two* distributive laws. Using the general notation for the operations, we may write the distributive laws as stating that for any three elements $a, b, c \in S$, we have,

$$a(b + c) = ab + ac$$
$$(b + c)a = ba + ca.$$

If we have a binary operation on a set, then it is natural to ask if it forms a group. In Mathematics, an important structure with two binary operations is the *ring* which is defined next.

Definition 5.1 *A system $[R, +, \cdot]$ of a set with two binary operations is called a* ring *if it is a commutative group with respect to addition, closed and associative with respect to multiplication, and the distributive laws hold.*

The additive identity in a ring is called *zero* and is written 0. The additive inverse is indicated by *minus* or $-$ and the shorthand $a - b$ is used for $a + (-b)$.

5.1.3 Examples of rings

As you saw in Activity 1, all of the systems that are familiar to you satisfy the distributive law, provided that the two operations are kept in the right order. Moreover, all of the familiar arithmetic systems form rings.

Proposition 5.1 *The systems $\mathcal{Z}$, $\mathcal{Q}$, $\mathcal{R}$, and $\mathcal{C}$ with ordinary addition and multiplication form rings.*

Whatever remains (after the activities) to prove this proposition is left for the exercises.

Some of the new systems studied in this book are also rings.

Proposition 5.2 *Each of the systems $[\mathcal{Z}_n, +_n, \cdot_n]$, $n = 2, 3, \ldots$ forms a ring.*

Proof. We know from Section 2.2.2, p. 60, that $[\mathcal{Z}_n, +_n]$ is a commutative group, that $\mathcal{Z}_n$ is closed with respect to multiplication mod n and that multiplication mod n is associative. It remains to establish distributivity and this is done "by inheritence" from $\mathcal{Z}$ analogously to the proofs of associativity of modular arithmetic.

$\square$

5.1.4 Rings with additional properties

Integral domains

Definition 5.2 *If the multiplication operation in a ring is commutative, then the ring is called a* commutative ring.

If the operation of multiplication has an identity, then it is called one *or* unity *and is written 1. The ring is then called a* ring with unity.

An element of a ring which has an inverse with respect to multiplication is called a unit.

Note that in the above definition, giving a name to an identity requires not only that it exists, but also that it is unique. In fact, this follows from the definition by an argument that should be familiar to you.

One of the most fundamental properties of arithmetic is that if you multiply two non-zero numbers, you never get zero for the answer. An interesting phenomenon that occurs with general systems of two binary operations is that this rule can be violated.

For example, in the ring $[\mathcal{Z}_6, +_6, \cdot_6]$, we have $2 \cdot_6 3 = 0$. On the other hand, there are some systems in which the usual rule *does* hold. Can you see that, in $[\mathcal{Z}_5, +_5, \cdot_5]$, if you multiply two non-zero elements (that is, 1, 2, 3, or 4) then you cannot get 0 (that is, a multiple of 5)?

In mathematics, we deal with such situations by formalizing the property and giving it a name. This helps us organize our study of when the rule does or does not hold, and what are its implications. In this spirit we make the following definition.

Definition 5.3 *A non-zero element a in a ring S is said to be a* zero divisor *if there is a non-zero element $b \in S$ such that either $ab = 0$ or $ba = 0$.*

The systems $[\mathcal{Z}_n, +_n, \cdot_n]$, $n = 2, 3, \ldots$ are particularly interesting in this connection. Some of them have zero divisors and some do not. Moreover, there is a simple condition that distinguishes between the two possibilities. Can you figure it out?

Take a look at the examples you constructed in Activity 10. Do these have zero divisors? What about the examples in Exercise 22?

One reason that zero divisors are important is that, in the next chapter, we want to study factorization in a general system. If you think about it, you might agree that the idea of factorization is not very useful if zero divisors are present. Our study of factorization will require their absence.

Definition 5.4 *A commutative ring with unity and no zero divisors is called an* integral domain.

Again, all of the familiar arithmetic systems form integral domains.

Proposition 5.3 *The systems $\mathcal{Z}$, $\mathcal{Q}$, $\mathcal{R}$, and $\mathcal{C}$ with ordinary addition and multiplication form integral domains.*

As we have seen, not all of the rings $\mathcal{Z}_n$ are integral domains. Have you figured out yet exactly which ones are?

Fields

If R is a ring, but you can't always find inverses, then given $a, b \in R$ with $a \neq 0$ you might not be able to solve the equation $ax = b$ for x. Our last system, which is the least general but has the most structure, is one in which such an equation can always be solved. As you will see, this turns out to open up the possibility of solving a lot more equations.

Definition 5.5 (integral domain.) *A commutative ring with 1 which has at least two elements and in which every non-zero element has an inverse is called a* field.

An equivalent form of this definition is that the system $(R, +, \circ)$ is a field if both $(R, +)$ and $(R - \{0\}, \circ)$ are Abelian groups, $1 \neq 0$ and the distributive property holds.

Now it turns out that some of the standard arithmetic systems are *not* fields. Do you see that $\mathcal{Q}$, $\mathcal{R}$, and $\mathcal{C}$ are fields, but $\mathcal{Z}$ is not? What about the $\mathcal{Z}_n$'s?

We have defined a field as a ring with certain additional properties. As it turns out, these properties imply the axioms for an integral domain.

Proposition 5.4 *A field is an integral domain.*

Proof. By comparing axioms we see that it is only necessary to show that a field cannot have zero divisors. Thus we assume $ab = 0$ and will show that a or b must be 0. Indeed, if $a \neq 0$, then multiplying by the inverse of

a it follows that $b = 0$. Similarly, if $b \neq 0$, we multiply by the inverse of b to get $b = 0$.

$\square$

Proposition 5.5 (Cancellation in an integral domain) *If R is an integral domain, then it satisfies the cancellation property:*

$$a \neq 0 \text{ and } a \circ b = a \circ c \text{ implies } b = c.$$

Proof. The main point of the proof is how the special property of integral domains (no zero divisors) is used. We will take for granted all the arithmetical properties that hold in any ring, such as those appearing in Activity 11 (and are true).

From the fact that $ab = ac$, it follows (since R is a group with respect to addition) that $ab - ac = 0$ and so $a(b - c) = 0$. Since R has no zero divisors and $a = 0$ is ruled out by assumption, it follows that $b - c = 0$, hence $b = c$.

$\square$

You should have a number of examples at hand which are integral domains, but not fields. All of them, however, must be infinite. In the finite case, the converse of Prop. 5.4 holds as well.

Proposition 5.6 *An integral domain with a finite number of elements is a field.*

Proof. An integral domain R is a commutative ring with 1 and so it is only necessary to show that every non-zero element has an inverse.

Let a be any non-zero element of R. Considering all of the powers, $a, a^2, a^3, \cdots$, it follows from the finiteness that there is, eventually, a repetition. That is, there exist positive integers $i > j$ with $a^i = a^j$, which can also be written as $a^j \cdot a^{i-j} = a^j \cdot 1$. Cancelling a^j, (why must a^j be non-zero?) we get $a^{i-j} = 1$, and so a^{i-j-1} is the inverse of a.

$\square$

5.1.5 *Constructing new rings from old—matrices*

In this and the following two subsections, we introduce constructions that will allow you to make your repertoire of examples of rings much richer than it is now. At the moment, the only rings that have been considered are $\mathcal{Z}, \mathcal{Q}, \mathcal{R}, \mathcal{C}$ and $\mathcal{Z}_n$, $n = 2, 3, \ldots$ Each of these rings can be used to make many new rings by forming matrices, polynomials, and functions. You have already worked with the first of these in Activity 10.

One of the values of having a diverse collection of examples at your fingertips is that potential theorems can be investigated. If you suspect that a general relationship is true, try to check it out on some of these examples. If you come up with an example in which your conjecture is false, then that is a counterexample, and you don't need to go further. If it seems to hold in several different kinds of examples, then not only does that strengthen your faith that it holds in general, but you might see, by working it out in an example, how a general proof of your conjecture might go.

In Activity 10 you were given a ring R and you constructed what is called the ring of 2×2 matrices over R. What this means is that, given the ring R a new ring $M(R)$ is formed as follows. The set consists of all 2×2 matrices whose entries are elements of R. Two matrices are considered equal if their corresponding entries are equal.

The addition is "matrix addition", which means that, given two matrices, you form a new matrix, each of whose elements is obtained by adding the corresponding entries in the given two matrices. The multiplication is standard matrix multiplication.

For example, here is the sum of two 2×2 matrices,

$$\begin{pmatrix} 1 & 2 \\ 3 & -1 \end{pmatrix} + \begin{pmatrix} 4 & -2 \\ 1 & 3 \end{pmatrix} = \begin{pmatrix} 1+4 & 2-2 \\ 3+1 & -1+3 \end{pmatrix} = \begin{pmatrix} 5 & 0 \\ 4 & 2 \end{pmatrix}$$

and here is their product,

$$\begin{pmatrix} 1 & 2 \\ 3 & -1 \end{pmatrix} \cdot \begin{pmatrix} 4 & -2 \\ 1 & 3 \end{pmatrix} = \begin{pmatrix} 1 \cdot 4 + 2 \cdot 1 & 1 \cdot (-2) + 2 \cdot 3 \\ 3 \cdot 4 + (-1) \cdot 1 & 3 \cdot (-2) + (-1) \cdot 3 \end{pmatrix}$$
$$= \begin{pmatrix} 6 & 4 \\ 11 & -9 \end{pmatrix}.$$

It would be helpful for you to write out the formulas for addition and multiplication in $M(R)$ for general matrices, using notation like

$$A = \begin{pmatrix} a_{11} & a_{12} \\ a_{21} & a_{22} \end{pmatrix}.$$

The ring properties are easy to check, but tedious. For example, $M(R)$ is a commutative group with respect to addition, roughly because the various properties in R are just transferred to each entry of the matrix. This is clearly true of closure, associativity, and commutativity. For the additive identity, you just take the identity, 0, in R and put it at each entry. Similarly for inverses.

The multiplication is a little more complicated (but there is less to check). Clearly, the entries in the product will still be in R because nothing but ring operations are performed on them. For the associative and distributive properties, you have to write out some formulas. You will have a chance to do this in the exercises.

An interesting question that arises with the ring $M(R)$ has to do with some of the additional properties we have discussed. Consider commutativity. When is the ring $M(R)$ commutative? Does it help to assume that R is a commutative ring?

If R has no zero divisors, what about $M(R)$? Can you find two non-zero matrices whose product is the zero matrix? What about units? If R has a multiplicative identity, what is the identity matrix? What will be the units in $M(R)$?

If R is an integral domain or a field, is the same true of $M(R)$? Would anything change if, instead of considering all of $M(R)$, you looked at a subset, say the set of invertible 2×2 matrices. Is anything different?

We collect some of the facts in the following proposition.

Proposition 5.7 *Let R be a ring.*

1. *$M(R)$ is a ring.*

2. *If R has more than one element, then $M(R)$ is not commutative.*

3. *If R is a ring with unity, then so is $M(R)$.*

4. *If R has more than one element, then $M(R)$ has zero divisors.*

It is not hard to see how to prove the statements in this proposition. In the second, for example, you need only find two matrices whose product is different if you reverse the order. Try it. You may find it easier than you expected.

Once you have a unity 1 in R it should be easy to check that the matrix

$$A = \begin{pmatrix} 1 & 0 \\ 0 & 1 \end{pmatrix}$$

is an identity for $M(R)$.

For zero divisors, just try to write down two non-zero matrices with as many as possible of the four entries equal to zero. For example, you can have all but one of the entries of the matrix equal to zero. You only have to choose the positions for the non-zero entries carefully.

The ring $M(R)$ we have just constructed has been called the ring of 2×2 matrices over R. In a similar way, the ring of $n \times n$ matrices over R can be constructed and its properties (analogous to Proposition 5.7) proved.

5.1.6 Constructing new rings from old—polynomials

Let R be a ring and x a symbol which has no particular meaning. It is possible to form expressions (analogous to polynomials with numerical coefficients) involving the elements of R, the symbol x and formal addition and multiplication. Thus, if $R = \mathcal{Z}$, one might write,

$$x, \quad 3x, \quad 3x^2, \quad 3x + 7x^5, \quad 1 - 2x - 2x^3 + x^4$$

and, in general,

$$a_0 + a_1 x + \cdots + a_{n-1} x^{n-1} + a_n x^n$$

where $a_0, a_1, \ldots, a_{n-1}, a_n$ are elements of R.

Such an expression is called a *polynomial* over R, and the ring elements $a_0, a_1, \ldots, a_{n-1}, a_n$ are called its *coefficients*. If $a_n \neq 0$, then n is called the *degree* of this polynomial and a_n its *leading coefficient*. The degree of a polynomial f is denoted $\deg(f)$. We denote by $R[x]$ the set of all polynomials in x with coefficients in R.

Two polynomials in $R[x]$ are considered equal if equal powers have equal coefficients. More precisely, suppose $f, g \in R[x]$ are given by

$$f(x) = a_n x^n + a_{n-1} x^{n-1} + \cdots a_1 x + a_0$$

$$g(x) = b_m x^m + b_{m-1} x^{m-1} + \cdots b_1 x + b_0.$$

Then, by adding zero terms if necessary, we can always rewrite them as

$$f(x) = a_k x^k + a_{k-1} x^{k-1} + \cdots a_1 x + a_0$$

$$g(x) = b_k x^k + b_{k-1} x^{k-1} + \cdots b_1 x + b_0.$$

where $k = \max(n, m)$. We now define $f(x)$ and $g(x)$ to be equal if

$$a_0 = b_0, a_1 = b_1, \ldots, a_k = b_k.$$

Addition and multiplication can be defined in $R[x]$ by transferring the operations from R and by imitating ordinary addition and multiplication of polynomials.

For example, consider the two polynomials $f(x) = 3x + 7x^5$ and $g(x) = 1 - 2x - 2x^3 + x^4$ in $Z[x]$. Their sum is given by

$$f(x) + g(x) = 3x + 7x^5 + 1 - 2x - 2x^3 + x^4 = 1 + x - 2x^3 + x^4 + 7x^5.$$

In other words, you just add the coefficients of "like terms", that is, terms with the same power of x.

In general, if $f(x), g(x)$ are polynomials in $R[x]$ then we can rewrite them (by adding zero terms as above) as

$$f(x) = a_k x^k + a_{k-1} x^{k-1} + \cdots a_1 x + a_0$$

$$g(x) = b_k x^k + b_{k-1} x^{k-1} + \cdots b_1 x + b_0.$$

Then the polynomial $f(x) + g(x)$ is defined by

$$f(x) + g(x) = (a_k + b_k) x^k + (a_{k-1} + b_{k-1}) x^{k-1} + \cdots (a_1 + b_1) x + (a_0 + b_0).$$

For the product, consider again the two polynomials, $f(x) = 3x + 7x^5$ and $g(x) = 1 - 2x - 2x^3 + x^4$ in $\mathcal{Z}[x]$. Their product is obtained as follows.

$$
\begin{aligned}
f(x)g(x) &= (3x + 7x^5)(1 - 2x - 2x^3 + x^4) \\
&= 3x - 6x^2 - 6x^4 + 3x^5 + 7x^5 - 14x^6 - 14x^8 + 7x^9 \\
&= 3x - 6x^2 - 6x^4 + 10x^5 - 14x^6 - 14x^8 + 7x^9
\end{aligned}
$$

In other words, you form all products (using distributivity) and then collect like terms.

In general, if $f(x), g(x) \in R[x]$ with

$$
f(x) = a_n x^n + a_{n-1} x^{n-1} + \cdots a_1 x + a_0
$$

$$
g(x) = b_m x^m + b_{m-1} x^{m-1} + \cdots b_1 x + b_0
$$

then the polynomial $f(x)g(x)$ is defined as follows.

$$
f(x) \cdot g(x) = c_{n+m} x^{n+m} + c_{n+m-1} x^{n+m-1} + \cdots c_1 x + c_0
$$

where,

$$
c_j = \sum_{i=0}^{j} a_{j-i} b_i = a_j b_0 + a_{j-1} b_1 \cdots a_1 b_{j-1} + a_0 b_j, \quad j = 0, 1, 2, \ldots m+n
$$

It is not difficult to check that $R[x]$ will always be a ring and again there are interesting questions about the extent to which properties of $R[x]$ are "inherited" from R.

Proposition 5.8 *Let R be a ring.*

1. *$R[x]$ is a ring.*

2. *If R is commutative, then so is $R[x]$.*

3. *If R is a ring with unity, then so is $R[x]$.*

4. *For $f, g \in R[x]$, $\deg(fg) \leq \deg(f) + \deg(g)$*

5. *If R is an integral domain, then so is $R[x]$. In this case, if $f, g \in R[x]$ we have the following formula for degrees,*

$$
deg(fg) = deg(f) + deg(g)
$$

6. *$R[x]$ is never a field.*

Proof. The first three points are straightforward. The commutativity of multiplication follows directly from manipulation of the formulas. In general, elements of R appear in $R[x]$ as polynomials of degree 0. In particular, the additive identity in $R[x]$ is 0 and, given a multiplicative identity 1 for R, it should be clear to you that the identity for $R[x]$ is the polynomial 1.

The other two statements follow from consideration of degrees and leading terms. Thus if f and g are non-zero polynomials with leading coefficients $a_n \neq 0$ and $b_m \neq 0$, respectively, then their product will have the form

$$fg(x) = a_n b_m x^{n+m} + \text{ terms of lower degree.}$$

Thus, $\deg(fg) \leq n + m = \deg(f) + \deg(g)$.

In particular, if R is an integral domain, then $a_n b_m \neq 0$ hence $fg \neq 0$ and its degree is $n + m$. Thus $R[x]$ is an integral domain and 5 is proved. It also leads to a proof of the sixth statement. Indeed, if R has zero divisors, then so has $R[x]$ and hence it is not a field. Therefore we may assume that R has no zero divisors. In this case, you cannot multiply a polynomial of positive degree by anything (other than 0) that will yield a polynomial of lower degree, in particular, you will never get 1. Hence $R[x]$ cannot be a field.

$$\square$$

5.1.7   Constructing new rings from old—functions

Our last general method of constructing rings makes use of functions. Let R be any ring and A any set. We denote by $F(A, R)$ the set of all functions whose domain is A and whose values lie in R. The operations in R transfer to $F(A, R)$ via "pointwise addition and multipication". That is, if $f, g \in F(A, R)$, then $f + g$ and $f \cdot g$ are defined by setting, for each $x \in A$,

$$(f + g)(x) = f(x) + g(x)$$

$$(f \cdot g)(x) = f(x) \cdot g(x)$$

Once again, the ring properties are easily checked and the questions about $F(A, R)$ being an integral domain or field arise.

Proposition 5.9 *Let R be any ring and A any set.*

1. *$F(A, R)$ is a ring.*

2. *If R is commutative, then so is $F(A, R)$.*

3. *If R is a ring with unity, then so is $F(A, R)$.*

4. *If A has more than one element, then $F(A, R)$ always has zero divisors.*

Proof. Exercise 21.

The most interesting point in this collection of facts is the last. Can you see how two non-zero functions f, g can multiply to yield the zero function?

The idea is to construct f and g so that they have non-zero values for different points in the domain.

The most important special case of this construction of rings of functions is the case when $R = \mathcal{R}$, the field of real numbers, and $A \subset \mathcal{R}$. Thus we are talking here about the ring of all functions from a (fixed) subset of the reals into the reals. This ring is of considerable interest in the branch of mathematics called analysis, especially when A is the unit interval or all of $\mathcal{R}$. The ring of functions $F(A, \mathcal{R})$ is important, but even more important is the subset of those functions which are continuous. What does it mean to assert that this subset is also a ring?

In the case when $A = \mathcal{R}$, do you see any connection between $F(\mathcal{R}, \mathcal{R})$ and $\mathcal{R}[x]$?

5.1.8 Elementary properties—arithmetic

The properties in Activity 11, at least those that are true, specify the extent to which you can do ordinary arithmetic in an arbitrary ring. We don't want to spoil your fun by proving them for you, but, just as an example, here is one way to prove the statement in Activity 11(a), that in any ring, multiplication by 0 results in 0.

Let a be an arbitrary element of a ring. Then, using the property of an identity (since a ring is an additive group) and the distributive law,

$$a \circ 0 = a \circ (0 + 0) = a \circ 0 + a \circ 0$$

and so, by cancellation (with respect to addition) it follows that

$$a \circ 0 = 0$$

The other equality in 11(a) is proved similarly.

Not everything is sure to work in a ring. Consider the statement in Activity 11(h). Can you see that this is true in $\mathcal{Z}_5$ but not in $\mathcal{Z}_6$? In this connection, recall Proposition 5.5, p. 159.

5.1.9 EXERCISES

1. What happens to the distributive laws if you interchange "addition" and "multiplication"? Find an example of a set and a pair of binary operations such that the distributive laws hold one way but not the other. Can you find an example in which the distributive laws hold both ways?

2. Check that $\mathcal{Z}$, $\mathcal{Q}$, $\mathcal{R}$, and $\mathcal{C}$ with ordinary addition and multiplication form rings (Proposition 5.1, p. 157).

3. Prove that the set $\mathcal{Z}[\sqrt{2}] = \{a + b\sqrt{2} : a, b \in \mathcal{Z}\}$ is an integral domain relative to ordinary addition and multiplication of real numbers.

4. Prove that the set $\mathcal{Z}[i] = \{a + bi : a, b \in \mathcal{Z}\}$, where $i = \sqrt{-1}$, is an integral domain relative to addition and multiplication of complex numbers. ($\mathcal{Z}[i]$ is called the ring of *Gaussian integers* and plays an important role in number theory.)

5. Suppose we replace $\mathcal{Z}$ with $\mathcal{Q}$ in the last two exercises. Show that the sets so obtained are fields. (Hint: Define the *conjugate* of $a + b\sqrt{2}$ by $a - b\sqrt{2}$ and of $a + bi$ by $a - bi$. In both systems check what happens when an element is multiplied by its conjugate.)

6. Complete the proof of Proposition 5.7, p. 161 concerning the matrix ring $M(R)$.

7. Prove that the set of all invertible elements in a ring R with 1 forms a multiplicative group. (These elements are called *units* in the ring R, and this group is denoted $U(R)$.)

8. Determine the group $U(R)$ for the following rings R: $\mathcal{Z}$, $\mathcal{Z}_{12}$, $\mathcal{Z}_7$, $\mathcal{Z}[\sqrt{2}]$, $\mathcal{Z}[i]$, $F[x]$ where F is a field.

9. Determine the group $U(M(F))$, where F is a field. In particular, show that the order of this group, for any field, is at least 6. (Recall that, by definition, a field has at least 2 distinct elements).

10. Show that, in an arbitrary field, $0 \neq 1$.

11. Find an example of a ring with an element a satisfying $a^2 = 0$. Do the same for $a^3 = 0$. Can you find for each n a ring element satisfying $a^n = 0$?

12. Suppose that a is an element in a ring with 1, and that $a^2 = 0$. Prove that $1 - a$ is a unit.

13. Suppose that a is an element in a ring with 1 and $a^n = 0$ for some positive integer n. Prove that $1 - a$ is a unit. (Hint: Recall that $1/(1 - a)$ is the sum of a decreasing infinite geometric series. Use this to *guess* the inverse of $1 - a$, then prove your guess.)

14. Prove the following properties for any ring R.

    (a) For all a in R, $a0 = 0a = 0$.

    (b) For all a, b in R, $a(-b) = (-a)b = -(ab)$.

    (c) For all a, b in R, $(-a)(-b) = ab$.

    (d) For all a, b, c in R, $a(b - c) = ab - ac$ and $(b - c)a = ba - ca$.

    (e) If R has 1, then for all a in R, $(-1)a = a(-1) = -a$.

(f) If R is commutative then for all a, b in R, $(a+b)^2 = a^2 + 2ab + b^2$.

(g) If R is commutative then for all a, b in R, $(ab)^2 = a^2b^2$.

(h) Give examples to show that if R is not commutative then the last two properties are not in general true.

15. For any two rings A and B define $A \oplus B = \{(a,b) : a \in A, b \in B\}$. Show that it is possible to define (pointwise) operations on $A \oplus B$ so that it becomes a ring. (This ring is called the *direct sum* of A and B.) If A has n elements and B has m elements, how many elements does $A \oplus B$ have?

16. Give an example of a ring R with unity 1 and a subring S with unity $1'$, such that $1' \neq 1$ and $1' \neq 0$. (Recall that in groups such a situation is impossible. See Proposition 2.10, p.52.)

17. Show that a commutative ring with unity is an integral domain if and only if it has the multiplicative cancellation property.

18. Suppose a, b are elements in an integral domain, satisfying $a^4 = b^4$ and $a^7 = b^7$. Prove that $a = b$. Generalize.

19. Let $R = F(A, \mathcal{R})$ be the ring of real-valued functions as defined in Section 5.1.7. Show that every non-zero element of R is either a unit or a zero-divisor. Is this true for *any* ring?

20. A ring R is called *Boolean* if $a^2 = a$ for all $a \in R$. Prove that a Boolean ring must be commutative.

21. Prove Proposition 5.9, p. 164.

22. For this problem, you have to use only **VISETL**.

Write a **proc** **polring** which will assume that **name_ring** has been run. Your **proc** will construct from R a new set **polR** and from **ad** and **mul**, two binary operations, **polad** and **polmul**. The set will consist of all **tuples** of any length, each of whose elements is an element of R. These **tuples** are to be considered as polynomials with coefficients coming from R. The addition will be term-by-term addition of coefficients and the multiplication will be multiplication of polynomials.

Apply **polring** to all of the examples in Activity 1, p. 153 and then apply is_ring, is_integral_domain, and is_field to the result. What can you say about "preservation"? That is, to what extent is it the case that if one of these **funcs** returns **true** for the original system, then it and/or the others will return **true** for the result of **polring**.

Explain why you cannot use **ISETL** in this problem, even when R is finite.

5.2 Ideals

5.2.1 ACTIVITIES

1. Write an **ISETL** func is_subring that assumes name_ring has been run and accepts a subset S of the ring R. Your func will determine if the set S is a subgroup of R (with respect to addition) and if, moreover, the set S is closed with respect to multiplication.

   Test your func on the following examples. Use **ISETL** where appropriate and otherwise use **VISETL**.

   (a) The ring is $\mathcal{Z}_{12}$ and S is, in turn, each of the subgroups (with respect to addition) of $\mathcal{Z}_{12}$. (See Theorem 3.3, p. 100.)

   (b) The ring is $M(\mathcal{Z}_2)$, the ring of 2×2 matrices with entries from $\mathcal{Z}_2$, and the set S is the subset of those matrices which have inverses.

   (c) The ring is $M(\mathcal{Z}_2)$, the ring of 2×2 matrices with entries from $\mathcal{Z}_2$, and the set S is the subset of those matrices which are triangular.

   (d) The ring is $M(\mathcal{Z}_2)$, the ring of 2×2 matrices with entries from $\mathcal{Z}_2$, and the set S is the subset of those matrices which are diagonal.

   (e) The ring is $M(\mathcal{Z}_2)$, the ring of 2×2 matrices with entries from $\mathcal{Z}_2$, and the set S is the subset of those matrices which are scalar. (A *scalar matrix* is a diagonal matrix in which all diagonal entries are equal.)

   (f) The ring is $M(\mathcal{Z}_4)$, the ring of 2×2 matrices with entries from $\mathcal{Z}_4$, and the set S is the subset of those matrices in which every entry is either 0 or 2.

   (g) The ring is $\mathcal{Z}_4[x]$, the ring of polynomials with coefficients in $\mathcal{Z}_4$, and the set S is the subset of those polynomials of degree ≤ 3.

   (h) The ring is $\mathcal{Z}_4[x]$, the ring of polynomials with coefficients in $\mathcal{Z}_4$, and the set S is the subset of those polynomials in which the constant term (i.e., the coefficient of x^0) is 0.

   (i) The ring is $\mathcal{Z}_4[x]$, the ring of polynomials with coefficients in $\mathcal{Z}_4$, and the set S is the subset of those polynomials in which the constant term and the term of degree 1 are both 0.

   (j) The ring is $\mathcal{Z}_4[x]$, the ring of polynomials with coefficients in $\mathcal{Z}_4$, and the set S is the subset of those polynomials in which each coefficient is either 0 or 2.

2. Update your func name_ring in a manner analogous to the version of name_group that provides for a subgroup as an optional argument,

constructs cosets (with respect to addition), and the operation of addition extended to cosets (see p. 110). Introduce a second operation on cosets by extending the ring multiplication to cosets as follows. Given two cosets A, B apply the **ISETL** operation arb to each, obtaining elements a, b of A, B respectively. The result of coset multiplication is then taken to be the coset of ab.

Thus, the result is the set of cosets (with respect to addition) and two computer operations — coset addition and coset multiplication.

Apply your operation of coset multiplication to cosets taken from examples in Activity 1 for which is_subring returns true. Is it always the case that applying coset multiplication to the same two cosets twice will always give the same coset as the answer?

Explain why the resulting set of cosets is sure to be a group with respect to addition.

3. Consider each of those examples of Activity 1 for which is_subring returned true and, for each of them, run name_ring and then apply is_ring to the set of cosets with coset addition and multiplication.

   Note that, in view of Activity 2 this activity can be done much more quickly than by actually running all of is_ring.

   For which of the examples does is_ring applied to the set of cosets return true?

4. Write an **ISETL** func is_ideal that assumes name_ring has been run with a subset for which is_subring returns true, so that you have a ring R and a subring S. Your func is to determine if the following condition holds.

$$\text{For all } r \in R, s \in S, \quad sr \in S \text{ and } rs \in S$$

   Run your func is_ideal on all of the examples of Activity 1 for which is_subring returned true.

5. Compare your results from Activities 3 and 4. Try to formulate some general observations and explain why they are true.

6. In this activity you are to consider all of the examples in Activity 3 for which is_ring applied to the set of cosets, with coset addition and multiplication, returned true. When this is the case, refer to the ring as R, the subring as I and the set of cosets as RmodI. Use appropriate funcs and/or **VISETL** to answer the following questions.

   (a) For which examples are R, RmodI commutative?

   (b) For which examples can you find elements a, b in R but not in I for which ab is in I?

(c) For which examples can you find another subset J, containing every element of I but different from I and R, for which is_ideal returns **true**?

(d) For which examples is RmodI an integral domain?

(e) Are there any conditions on I which guarantee that RmodI is an integral domain?

(f) For which examples is RmodI a field?

(g) Are there any conditions on I which guarantee RmodI is a field?

5.2.2   Analogies between groups and rings

A great deal of the study of rings is analogous to the study of groups. This is because the ring structure $[R, +, \circ]$ includes a part which is a commutative group. To see this we need only "forget" the multiplication operation of the ring, and what we are left with is a set and an addition, $[R, +]$, which satisfies all the axioms for a commutative group. We will therefore use mostly additive notation for the group operation in a ring. For a capsule summary of the relevant group properties, we recall that certain subsets of a group are subgroups; these are used to form cosets and the product (or, in the present context, addition) of cosets is defined; in certain cases, when the subgroup is normal (which for the present purpose is always the case since the group is commutative), this coset product leads to the formation of a new group, called the quotient group; homomorphisms between groups are maps that preserve the group operation; there is a close connection between normal subgroups and homomorphisms arising out of the fact that the kernel of a homomorphism is a normal subgroup; the converse is also true, that is, given a normal subgroup N of a group G, we can find a homomorphism (namely the canonical homomorphism onto the quotient group G/N) with N as its kernel; and finally, the fundamental homomorphism theorem characterizes all homomorphic images of a group as being (isomorphic to) its quotient groups.

One can try to repeat this program, step-by-step for rings. Of course, it will always be necessary to check things like closure and preservation of operation for both addition and multiplication but, otherwise, one can at least try to just transfer everything from groups to rings.

The amazing thing is that to actually carry out this plan we need introduce no new ideas — it works in a quite straightfoward fashion — with one exception. The exception, which you had the opportunity to explore in Activities 2–5, is the condition that makes coset operations work. There is no problem with addition of cosets because $[R, +]$ is a commutative group so that all its subgroups are automatically normal. The one change needed in passing from groups to rings is that a new condition is required for coset multiplication to work. This leads to the notion of an *ideal*.

We now proceed to run through this program.

5.2.3 Subrings

Definition of subring

A subgroup of a group G is a subset S of G such that the group operation, restricted to S, makes S a group. We have the same thing for rings.

Definition 5.6 (subring.) *Let $[R, +, \circ]$ be a ring and let S be a subset of R. If $[S, +, \circ]$ is a ring then we say that S is a* subring *of G.*

Note. Strictly speaking the operations on S are not quite the same $+, \circ$, but rather their *restrictions* to S.

Conceptually, the func is_subring, that you wrote for Activity 1, just checks the group axioms for addition and closure and associativity for multiplication. In practice, this will take a lot more time and effort than necessary. A much shorter func can be constructed using the following test.

Proposition 5.10 (Test for subring.) *A subset S of a ring R is a subring if and only if the following conditions hold.*

1. *S is closed with respect to the addition and multiplication of R.*

2. *S contains 0 (the zero element of R).*

3. *The additive inverse (in R) of every element of S is in S.*

Proof. Compare Proposition 3.1, p. 87.

5.2.4 Examples of subrings

Subrings of $\mathcal{Z}_n$ and $\mathcal{Z}$

Most (but not all) of the examples in Activity 1 are subrings. Looking at Activity 1(a) you determined, hopefully, that every subgroup of $\mathcal{Z}_{12}$ is actually a subring. This is a general fact about numbers.

Theorem 5.1 (Subrings of $\mathcal{Z}_n$.) *Every (additive) subgroup of $\mathcal{Z}_n$ or of $\mathcal{Z}$ is a subring.*

Proof. One way to prove this would be to recall that the subgroups of $\mathcal{Z}_n$ are obtained from the factors k of n in that for each factorization $n = kd$, the subset $S = \{0, d, 2d, \dots, (k-1)d\}$ is a subgroup of $\mathcal{Z}_n$ isomorphic (as a group) to $\mathcal{Z}_k$ and this gives all of the subgroups of $\mathcal{Z}_n$ (Theorem 3.3, p. 100.) It is then straightforward to check that each of these subgroups is actually a subring.

There is a simpler and more elegant approach that uses the intimate relationship between addition and multiplication in $\mathcal{Z}_n$ — that is, multiplication is repeated addition. Thus, if A is a subgroup of the ring $\mathcal{Z}_n$ and a, b are in A, then

$$ab = b + \cdots + b$$

where there are a terms in the addition. Hence, by the subgroup property, $ab \in A$.

The case of $\mathcal{Z}$ is proved similarly.

$\square$

Here is a short list of some examples of subrings of $\mathcal{Q}$, $\mathcal{R}$, $\mathcal{C}$. (Can you find some more?)

1. $\mathcal{Z}$ is a subring of $\mathcal{Q}$.

2. $\mathcal{Q}$ is a subring of $\mathcal{R}$.

3. $\mathcal{R}$ is a subring of $\mathcal{C}$.

4. The set of Gaussian integers, $G = \{a + b\mathbf{i} : a, b \in \mathcal{Z}\}$ is a subring of $\mathcal{C}$.

Subrings of $M(R)$

In Activity 1(b) you saw that the set of invertible matrices is *not* a subring of $M(\mathcal{Z}_2)$. Can you see a general reason for this? Can you express it in a way that will show that no matter what ring R you begin with, the subset of invertible matrices will not be a subring?

On the other hand, the set of diagonal matrices and the set of scalar matrices *are* subrings. Again this is a general fact that is true for any ring of matrices $M(R)$. Try to prove it. What about the set of triangular matrices?

Finally, in Activity 1(f) there is a general method of constructing subrings. In this case, R is the ring $\mathcal{Z}_4$ and $A = \{0, 2\}$ is a subring of R. Then, in the set of matrices $M(R)$ we consider the subset S of those matrices whose entries are in the subring A of R. The subring property of A is "lifted" to S which is essentially the ring $M(A)$.

Subrings of polynomial rings

The subset of polynomials with degree less than a fixed number (as in Activity 1(g)) can fail to be a subring (except if this fixed number is 0) because the product of two polynomials of positive degree can have a degree which is the sum of the two degrees so the upper bound will eventually be exceeded. This difficulty will not always occur, however. Think about zero divisors and integral domains, and see if you can work out a general description of what can happen.

In any case, Activity 1(g) provides an example of a subgroup which is *not* a subring.

The other three examples in Activity 1(h), (i), and (j) are all subrings. The last one is analogous to the general construction used in M(R) (Activity 1(f)).

Subrings of rings of functions

You will recall (p. 164) that if $\mathcal{R}$ is the set of real numbers and $I \subset \mathcal{R}$ then $F(I, \mathcal{R})$ is a ring. Here is a way of constructing subrings that is important in analysis. Let $J \subset I \subset \mathcal{R}$. We can then take $F_J(I, \mathcal{R})$ to be the set of all functions f in $F(I, \mathcal{R})$ which vanish on J, that is, for which $f(x) = 0$, $\forall x \in J$. It is easy to check that this is a subring.

5.2.5 Ideals and quotient rings

Definition of ideal

The main purpose of Activities 2–5 was for you to think about, and possibly discover, the conditions under which coset addition and multiplication work nicely for a subring of a ring. There is no difficulty with addition since this operation is assumed, in a ring, to be commutative, so all additive subgroups, in particular those arising from subrings, are normal. The problem is coset multiplication and here there is an issue which must be resolved even before considering closure and associativity for this operation.

Consider a ring R and a subring S. If A, B, are cosets of S then we can take arbitrary elements ("representatives") a from A and b from B and it follows that

$$A = a + S, \quad B = b + S.$$

In Activity 2 the product AB of the two cosets was taken to be the coset $ab + S$. The issue is that the choice of "representatives" a and b from A and B respectively, is arbitrary: how can we be sure that if different choices, say a', b' in A, B are made then the resulting coset $a'b' + S$ is the same as $ab + S$? In other words, does this new procedure actually define an operation that takes two cosets and returns a definite coset?

In Activity 2 you should have seen some examples in which an operation is defined and some in which it is not. What we need is a condition that guarantees that coset multiplication is defined. Once it is defined, it is clear that the result will be a coset so there is no problem with closure, and associativity and distributivity will be "inherited" from the same properties in the ring R.

It turns out that the condition you worked with in Activity 4 does the entire job, so we formalize it as an important definition—the ring-theory analog of a normal subgroup.

Definition 5.7 (ideal.) *A subset I of a ring R is called an* ideal *if*

I *is an additive subgroup of* R

and

for all $r \in R, a \in I, ra \in I$ *and* $ar \in I$

Now we must prove that this does the job. Perhaps you observed in Activity 5 that the list of examples for which coset multiplication "works" (Activity 3) is the same as the list for which the ideal condition holds (Activity 4). The following theorem (in which we also incorporate the discussion on addition of cosets) establishes that this is true in general.

Theorem 5.2 *Let* I *be an ideal of the ring* R, *and let* R/I *be the set of additive cosets of* I *in* R. *Then the formulas for coset addition and multiplication "by representatives"*

$$(a + I) + (b + I) = (a + b) + I$$

$$(a + I)(b + I) = ab + I.$$

define binary operations on cosets (called coset addition and coset multiplication), and the set R/I *is a ring relative to these operations.*

Proof. As we discussed before, we already know that R/I is an Abelian group relative to coset addition since the additive group of R is Abelian, so the additive subgroup I is normal. What is left to prove, then, is that R/I satisfies the ring properties concerning multiplication, and chief among these is that multiplication is at all *defined*. Let A,B be cosets and choose $a \in A$, $b \in B$ so that $A = a + I$, $B = b + I$ (Proposition 3.17, p. 113). Suppose $a' \in A$ and $b' \in B$ so that, also, $A = a' + I$, $B = b' + I$. We must show that

$$ab + I = a'b' + I.$$

We can make use of the fact that $a = a' + x$ and $b = b' + y$, for some $x, y \in I$. Then we have

$$ab = (a' + x)(b' + y) = a'b' + a'y + xb' + xy.$$

Now all terms but the first are in I because of the ideal property; hence, if we set $z = a'y + xb' + xy$ then we have $ab = a'b' + z$ with $z \in I$. This shows that $ab + I = a'b' + I$, as required.

Having established that coset multiplication is defined, it is clear that the set of cosets is closed with respect to this operation. The associativity and distributivity are "inherited" from those properties in R. (Hopefully you are able by now to "translate" the idea of inheritance into a detailed proof.)

[]

Definition 5.8 (quotient ring.) *Let I be an ideal in a ring R. The ring of cosets mod I with coset addition and coset multiplication is called the quotient ring of R mod I, and is written R/I.*

We will discuss below some examples of ideals, and in the next section we will try to identify the quotient rings that are obtained from these examples.

Examples of ideals

As you saw in Activities 3 and 4, many, but not all of the examples in Activity 1 are ideals.

For $\mathcal{Z}_n$ and $\mathcal{Z}$ "everything" is an ideal. This is so, as in the case of subrings (see Theorem 5.1, p. 172), because multiplication is just repeated addition in these rings.

Theorem 5.3 *Every (additive) subgroup of $\mathcal{Z}_n$ and of $\mathcal{Z}$ is an ideal.*

Proof. Exercise 1.

If you multiply a diagonal matrix by an arbitrary matrix, the result may not be diagonal. (Can you find an example?) Hence, the subring of $M(R)$ consisting of diagonal matrices is not an ideal.

On the other hand, you can always get an ideal in $M(R)$ by taking an ideal I in R and considering those matrices in $M(R)$, all of whose entries come from I. It is easy to check that this always gives an ideal in $M(R)$ and this is exactly what was done in Activity 1(f).

Turning to polynomial rings, we see that this same trick of replacing R with an ideal was used in Activity 1(j). You can check that this gives an ideal.

What about Activities 1(h) and (i)? If you multiply an arbitrary polynomial by another polynomial in which the constant term (or both the terms of degree 0 and 1) has zero coefficient, what can you say about the result? Does this lead to an ideal?

Look at p. 173 where there is a list of some subrings of $\mathcal{Q}$, $\mathcal{R}$, and $\mathcal{C}$. Are they also ideals? These amount to very simple questions about properties of numbers.

Finally, we point out that the subring $F_J(I, \mathcal{R})$ (see p. 174) is an ideal in $F(I, \mathcal{R})$.

5.2.6 Elementary properties of ideals

Two examples of ideals are "trivial", that is, they are always present: the entire ring and the ring $\{0\}$ are clearly ideals. In general it is easy to see when a given ideal is not $\{0\}$ — you only have to look for a single non-zero element. It can be a little more difficult to see that an ideal is not the whole ring. Following is a simple test that works for rings with unity.

Proposition 5.11 *Let R be a ring with unity 1, I an ideal in R. Then $I = R$ if and only if $1 \in I$.*

Proof. Obviously if $R = I$ then $1 \in I$. Conversely, suppose $1 \in I$. Then, for any $r \in R$ it follows from the ideal property that $r = 1r \in I$. Hence I is all of R.

$\square$

Corollary 5.1 *If, in a ring with unity, an ideal contains an invertible element, then the ideal must be the whole ring.*

Proof. If the ideal contains an invertible element a, then, by the ideal property, $1 = a^{-1}a \in I$.

$\square$

5.2.7 Elementary properties of quotient rings

The following proposition provides two simple examples of ring properties that are inherited by a quotient ring. (Compare Activity 6(a).)

Proposition 5.12 *Let I be an ideal in the ring R.*
If R is commutative then so is the quotient ring R/I.
If R has an identity then so has the quotient ring R/I.

Proof. Exercise 21.

Quotient rings that are integral domains — prime ideals

When is a quotient ring an integral domain? Here we do not have a simple relation between a property of a ring and the corresponding property for a quotient ring. Indeed, it doesn't go in either direction. That is, we can find examples of rings which are integral domains and yet they have quotient rings which are not. Moreover, we can find a ring which is *not* an integral domain, but some of its quotient rings are. Put another way, it can be said that the formation of quotient rings can eliminate zero divisors when they are present as well as introduce them when they are not.

You should compare the following examples with what you found in Activity 6(c).

Example. Let $R = \mathcal{Z}_{12}$ and $I = \{0, 3, 6, 9\}$. Then, as we have seen, I is an ideal. Notice that 3 and 4 are zero divisors in R. We can check the possibility of zero divisors in R/I.

Suppose that neither of $a + I$, $b + I$ are the zero coset I, that is, $a, b \notin I$. Then could it be that $ab + I$ is I, that is $ab \in I$? Introducing the meaning of these quantities in this example, we have integers in $\mathcal{Z}_{12}$ neither of which is a multiple of 3. You can see that their product cannot then be a multiple of 3. Hence, $ab \notin I$. Thus R/I has no zero divisors.

Example. Let $R = \mathbb{Z}$ and let I be the subset of all multiples of 6. We know that R has no zero divisors and that I is an ideal. What about zero divisors in R/I? Can you see that neither $2 + I$ nor $3 + I$ is zero in R/I, but their product, $6 + I = I$ is zero? Hence, in this case, R/I does have zero divisors.

Look carefully at these two examples and try to see what is really making the difference. What is causing or preventing the existence of zero divisors in the quotient ring? Does this have anything to do with Activity 6(c)? What about Activity 6(d)?

Do you see that the point is about elements and their products being or not being in I? One thing is clear: because I is an ideal, if a or b (or both) are in I then their product must also be in I. What about the converse? If the product ab is in I, is it possible that both a and b are *not* in I? This is the real issue. If you think about the interpretation of forming a quotient mod I as "smashing I down to 0", then this is also the issue for being 0 in the quotient ring.

Now we formalize all this in the following definition and proposition.

Definition 5.9 *A* proper *ideal* I *in a ring* R *is called a* prime ideal *if whenever a product ab of elements of R is in I, then either a or b (or both) are in I.*

Proposition 5.13 *Let* R *be a commutative ring with unity,* I *an ideal in* R. *Then* R/I *is an integral domain if and only if* I *is a prime ideal.*

Proof. Exercise 23.

Quotient rings that are fields—maximal ideals

Analogous to the question of a quotient ring being an integral domain, we may ask when a quotient ring is a field. Did you draw any conclusions from the results you obtained in Activities 6(c) and 6(d)? This question is a little more complicated. If the ideal is the entire ring, then the quotient ring will only have one element (zero) and so it cannot be a field since a field must have at least two elements. This means, because of Corollary 5.1, p. 177, that a quotient ring will not be a field if the ideal has any invertible elements. Since invertible elements are what is needed to have a field, one can take some intuitive comfort from thinking about ideals which are as large as possible without being the whole ring. It means that the ideal "just misses" having invertible elements.

This notion turns out to be the complete story.

Definition 5.10 *Let* R *be a ring. An ideal* I *in* R *is said to be* maximal *if* *it is not equal to R and there is no ideal other than I or R which contains I.*

Lemma 5.1 *Let I be an ideal in a commutative ring R with an identity 1, and let $r \in R$. Then the set $J = \{rs + a : s \in R, a \in I\}$ is an ideal which contains I and r.*

Proof. Exercise.

Theorem 5.4 *Let R be a commutative ring with identity and I an ideal. Then the quotient ring R/I is a field if and only if I is a maximal ideal.*

Proof. Suppose that I is a maximal ideal and let $r + I$, $r \in R$ be a non-zero element of R/I. This means that $r \notin I$ and we have to show that $r + I$ has an inverse in R/I. Let $J = \{rs + a : s \in R, a \in I\}$. Then by the lemma, J is an ideal containing I and r, hence *properly* containing I. Since I is maximal, it follows that $J = R$. In particular, $1 \in J$. This means that we can find $s \in R$, $a \in I$ such that

$$1 = rs + a.$$

It then follows that $(r + I)(s + I) = 1 - a + I = 1 + I$ which shows that $s + I$ is a multiplicative inverse of $r + I$ in R/I.

Conversely, suppose R/I is a field and let J be an ideal which properly contains I. Then it is easy to check that the set $\overline{J} = \{j + I : j \in J\}$ is an ideal in R/I. Because J contains I properly, $\overline{J}$ is a non-zero ideal in the field R/I so it must equal R/I. In particular, $1 + I \in \overline{J}$, that is, $1 = j + i$ for some $i \in I, j \in J$. Since I is contained in J, it follows that $1 \in J$ and so $J = R$. Thus I is a maximal ideal.

$\square$

5.2.8 EXERCISES

1. Prove that every (additive) subgroup of $\mathcal{Z}_n$ and of $\mathcal{Z}$ is an ideal (Theorem 5.3, p. 176).

2. Are any of the subrings in the following chain ideals?

$$\mathcal{Z} \subseteq \mathcal{Q} \subseteq \mathcal{R} \subseteq \mathcal{C}.$$

3. Show that $\overline{A} = \{(a, 0_B) : a \in A\}$ and $\overline{B} = \{(0_A, b) : b \in B\}$ are ideals of $A \oplus B$ whose intersection is the zero-ideal of $A \oplus B$. ($A \oplus B$ is defined in Exercise 15, p. 167.)

4. Find in $\mathcal{Z} \oplus \mathcal{Z}$ a prime ideal which is not maximal.

5. Find all maximal ideals in $\mathcal{Z}_n$.

6. Let R be a commutative ring with unity, $a \in R$, and define

$$\langle a \rangle = \{ra : r \in R\}$$

Prove that $\langle a \rangle$ is an ideal of R which contains a and that it is the smallest such ideal; that is, if J is any ideal of R which contains a, then $I \subseteq J$. ($\langle a \rangle$ is called the ideal *generated* by a.)

7. If R is any ring, we still want $\langle a \rangle$ (the ideal generated by a) to be the smallest ideal of R which contains a. Will the set $\langle a \rangle$ defined in the previous exercise do? If not, fix it so that you can prove that the new set has the desired properties.

8. Let R be a commutative ring with unity, $a, b \in R$, and define

$$\langle a, b \rangle = \{ra + sb : r, s \in R\}$$

Prove that $\langle a, b \rangle$ is an ideal of R which contains a and b and that it is the smallest such ideal; that is, if J is any ideal of R which contains a and b, then $I \subseteq J$. ($\langle a, b \rangle$ is called the ideal *generated* by a and b.)

9. What is the ideal $\langle 4, 6 \rangle$ in $\mathcal{Z}$? Find a single number $a \in \mathcal{Z}$ that will generate the same ideal (that is, so that $\langle a \rangle = \langle 4, 6 \rangle$). Repeat the same question with $\langle 3, 5 \rangle$.

10. Can we find, for any $x, y \in \mathcal{Z}$, an $a \in \mathcal{Z}$ such that $\langle a \rangle = \langle x, y \rangle$?

11. Prove that the intersection of any number of ideals of R is again an ideal. Is the *union* of ideals of R necessarily an ideal?

12. If S is any subset of a ring R, show that the intersection of all ideals of R which contain S is the smallest ideal containing S. (It can thus serve as an alternative construction for the ideal generated by S).

13. Is the intersection of two prime ideals (see p. 178) a prime ideal?

14. If a, b are two integers, find an integer x such that $\langle x \rangle = \langle a \rangle \cap \langle b \rangle$.

15. State and prove a converse to Theorem 5.2, p. 175. That is, show that if A is a subring of a ring R such that "coset multiplication works", then A is an ideal.

16. Let $R = F(I, \mathcal{R})$ be the ring of functions as defined in Section 5.1.7, and let a be any point in the interval I. Prove that the set $J_a = \{f \in R : f(a) = 0\}$ is a maximal ideal of R. Can you guess what the field R/J_a looks like?

17. Let $P = \{a + bi : a, b \in \mathcal{Z}, a, b \text{ are both even or are both odd}\}$. Show

that P is a maximal ideal of $\mathcal{Z}[i]$, the ring of Gaussian integers. How many elements does $\mathcal{Z}[i]/P$ have?

18. Suppose R is a ring, I an ideal in R. Prove that $M(I)$ is an ideal of $M(R)$. (See p. 160 for a discussion of $M(R)$.)

19. Suppose R is a ring, J an ideal in $M(R)$. Prove that there exists an ideal I of R such that J is isomorphic to $M(I)$. Deduce that if F is a field, $M(F)$ has no non-trivial ideals. (Such a ring is called a *simple* ring.)

20. The last two exercises show that ideals of the matrix ring are precisely those that arise from ideals of the ground ring. Formulate and prove an analogous theorem for polynomial rings.

21. Show that the properties of commutativity and existence of an identity are inherited by quotient rings (Proposition 5.12, p. 177).

22. Prove that the prime ideals in $\mathcal{Z}$ are precisely those of the form $p\mathcal{Z}$, where p is a prime number.

23. Prove Proposition 5.13, p. 178: Let R be a commutative ring with unity, I an ideal in R. Then R/I is an integral domain if and only if I is a prime ideal.

24. Prove Lemma 5.1, p. 179 about the ideal "generated" by an ideal I and an element r.

25. Write the addition and multiplication tables for $2\mathcal{Z}/8\mathcal{Z}$. Is this ring isomorphic to $\mathcal{Z}_4$?

26. This exercise demonstrates a general method whereby quotient rings are used to remove an unwanted part of a ring (we annihilate this part by making it zero). An element in a ring is called *nilpotent* if some power of it is equal to 0. (For example, 2 is nilpotent in $\mathcal{Z}_8$.) The *radical* of a commutative ring R is defined as the set of all nilpotent elements of R, and is denoted $\mathrm{Rad}(R)$. Prove that if R is a commutative ring then $\mathrm{Rad}(R)$ is an ideal of R, and $\mathrm{Rad}(R/\mathrm{Rad}(R)) = \{0\}$ (that is, $R/\mathrm{Rad}(R)$ has no non-zero nilpotent elements).

27. Find the radical of $\mathcal{Z}_6$, $\mathcal{Z}_7$ and $\mathcal{Z}_{15}$.

28. Find all nilpotent elements in $M(\mathcal{Z})$. Do they form an ideal?

29. Show that the sets of all diagonal matrices (all items off the main diagonal are 0) and of all triangular matrices (all items below the main diagonal are 0) are subrings of the ring of all $n \times n$ matrices over a ring R.

30. Determine the smallest subring of $\mathcal{Q}$ which contains 1.

31. Determine the smallest subfield of $\mathcal{Q}$ which contains 1.

32. Determine the smallest subring of $\mathcal{Q}$ which contains $1/2$.

33. Let R be a commutative ring with unity, I an ideal of R. Consider the following three properties for I: principal, maximal, prime. Does any one of them imply the other? Prove or find a counterexample.

5.3 Homomorphisms and isomorphisms

5.3.1 ACTIVITIES

1. In Chapter 4, Section 4.2.1, you worked with the funcs is_hom and is_iso to look for homomorphisms and isomorphisms of groups. It might be reasonable to think of renaming those funcs is_ghom and is_giso respectively. Recall also (p. 129) that you used a second func name_group'.

   In analogy with the group situation, write funcs name_ring', is_rhom and is_riso.

   In Section 4.2.1, Activity 2 there is a list of maps, some of which turned out to be group homomorphisms. For those that were, and for which the groups can also be considered to be the additive group of some ring, use is_rhom to see if they are also ring homomorphisms.

   Using **VISETL** you can find a group isomorphism between Z and twoZ. Is this also a ring isomorphism?

2. In each of the examples from Activity 1 for which is_rhom returns true, find the kernel of the map. In which cases is it an ideal?

3. Let R be a commutative ring and $R[x]$ the ring of polynomials over R. Consider the map $T : R[x] \rightarrow R$ defined by: $T(f)$ is the constant term of f. Show that T is a ring homomorphism of $R[x]$ onto R.

   Find as many other homomorphisms of $R[x]$ onto R as you can. Does it make any difference to assume that R has 1?

4. How many rings can you find which have a subring isomorphic to $\mathcal{Z}$?

5. Same as the previous activity with $\mathcal{Z}$ replaced by $\mathcal{Z}_n$.

6. In this activity we present a list of quotient rings. For each quotient ring, try to find a "familiar" ring to which it is isomorphic.

   In most cases, the ring is constructed with the help of some auxiliary ring R. You can do the entire problem with **VISETL** or, you can replace R by some small ring such as $\mathcal{Z}_2, \mathcal{Z}_3$ or $\mathcal{Z}_4$, use **ISETL** to discover what is happening and then use your mind to generalize.

   Here is the list of quotient rings.

(a) The quotient ring $\mathcal{Z}_n/I$ for some $n = 2, 3, \ldots$, where I is an ideal of $\mathcal{Z}_n$.

(b) The quotient ring $\mathcal{Z}/I$ where I is an ideal of $\mathcal{Z}$.

(c) The quotient ring $M(R)/M(I)$ where R is a ring, I is an ideal in R, and $M(R)$, $M(I)$ are, respectively, the rings of 2×2 matrices with entries in R, I.

(d) The quotient ring $R[x]/I[x]$ where I is an ideal in R.

(e) The quotient ring $R[x]/I$ where R is a ring, $R[x]$ is the ring of polynomials over R and I is the ideal of polynomials in $R[x]$ whose constant term vanishes.

(f) The quotient ring $R[x]/I$ where R is a ring, $R[x]$ is the ring of polynomials over R and I is the ideal of polynomials in $R[x]$ whose terms of degree 0,1 vanish.

(g) The quotient ring $R[x]/I$ where R is a ring, $R[x]$ is the ring of polynomials over R and I is the ideal of polynomials in $R[x]$ whose terms of degree 0,1,2 vanish.

(h) The quotient ring $R[x]/I$ where R is a commutative ring with 1, $R[x]$ is the ring of polynomials over R, $f(x)$ is a fixed polynomial in $R[x]$, and I is the ideal of all polynomials in $R[x]$ of the form $f(x)g(x)$ for some $g(x) \in R[x]$ (i.e., the ideal *generated* by $f(x)$).

7. In Activity 2 you looked at ring homomorphisms $h : R \longrightarrow R'$ and their kernels I which were ideals. In each of these cases, form R/I and find a "familiar" ring to which it is isomorphic.

   Compare your results here with the results in Activity 6.

8. Find as many homomorphic images of $\mathcal{Z}$ as you can.

9. Find as many homomorphic images of $\mathcal{Z}_n$ as you can.

10. Find as many homomorphic images of $\mathcal{Z}_2[x]$ as you can.

5.3.2 Definition of homomorphism and isomorphism

The idea of a group homomorphism is that it is a map which "preserves" the group operation. An isomorphism is a homomorphism that is one-to-one and onto.

We can transfer this directly to rings by requiring that the *two* operations be preserved. The following formal definition should come as no surprise, especially after you have done Activity 1.

Definition 5.11 (ring homomorphism.) *Let R, R' be rings. A map $h : R \longrightarrow R'$ is a ring homomorphism if it preserves the two ring operations; that is, for all $a, b \in R$,*

$$f(a + b) = f(a) +' f(b)$$

$$f(a \circ b) = f(a) \circ' f(b).$$

A ring homomorphism is called a ring isomorphism *if it is one-to-one and onto.*

When the meaning is clear from the context, we will drop the modifier "ring" for homomorphisms and isomorphisms.

Group homomorphisms vs. ring homomorphisms

Clearly, a ring homomorphism is also a group homomorphism of the underlying additive groups of the two rings. Is the converse true? That is, if you have a map of one ring to another which is a homomorphism of the underlying additive groups, will it also be a ring homomorphism? In any case, the simplest place to look for ring homomorphisms is to consider two rings R, R' and a map $f : R \longrightarrow R'$ which is a homomorphism of the additive groups in these two rings. Will f also be a ring homomorphism?

Well, the answer is sometimes yes and sometimes no. Hopefully, you realized this in doing Activity 1 when you considered group homomorphisms from Chapter 4, Section 4.2. For example, the group homomorphism (Activity 2(c), p. 129) $h : \mathcal{Z}_{20} \longrightarrow \mathcal{Z}_5$ given by $h(x) = x \bmod 5$ is also a ring homomorphism. It is only necessary to verify the equalities:

$$\begin{aligned} h(xy) &= (xy \bmod 20) \bmod 5 = xy \bmod 5 \\ &= (x \bmod 5)(y \bmod 5) \bmod 5 = h(x)h(y) \end{aligned}$$

(Which of the above equalities depends on the fact that 5 divides 20?)

On the other hand, the group homomorphism (Activity 2(l), p. 129) $U : \mathcal{Z} \to \mathcal{Z}$ defined by $U(n) = 7n$ is not a ring homomorphism because, for example, $U(2 \cdot 3) = 7 \cdot 6 = 42$ while $U(2) \cdot U(3) = (7 \cdot 2) \cdot (7 \cdot 3) = 14 \cdot 21 \neq 42$.

Can you see another way of determining that the map U is not a homomorphism, by looking at the identity? We will see below that onto homomorphisms preserve identities. However 1 is a multiplicative identity in $\mathcal{Z}$ but $U(1) = 7$ is not an identity in $7\mathcal{Z}$ (which is the image of $\mathcal{Z}$ under U).

Similarly, the map that sends an integer to its double establishes that $\mathcal{Z}$ and $2\mathcal{Z}$ are isomorphic as groups. Do you see why they cannot be isomorphic as rings? (Note that it is not enough to show that this particular map fails.)

5.3.3   *Examples of homomorphisms and isomorphisms*

Homomorphisms from $\mathcal{Z}_n$ to $\mathcal{Z}_k$

In general, it is easy to check that for any positive integers n, k where k divides n, the map $h : \mathcal{Z}_n \longrightarrow \mathcal{Z}_k$ given by $h(x) = x \bmod k$ is a ring homomorphism and it is onto. We saw in Chapter 4, p. 149 that this formula gives all group homomorphisms of $\mathcal{Z}_n$, and since a ring homomorphism must be a group homomorphism, it follows that the formula gives all ring homomorphisms.

Furthermore, as we had in the case of groups (see p. 149), all homomorphisms of $\mathcal{Z}_n$ to a ring R' can be "factored" as $f \circ h$ where $h : \mathcal{Z}_n \longrightarrow \mathcal{Z}_k$ (for some divisor k of n) is the homomorphism of $\mathcal{Z}_n$ onto $\mathcal{Z}_k$ given by $h(x) = x \bmod k$, and f is a one-to-one homomorphism of $\mathcal{Z}_k$ into the ring R'.

Homomorphisms of $\mathcal{Z}$

A complete analysis of the homomorphisms of $\mathcal{Z}$ can be made in analogy with the preceeding analysis for $\mathcal{Z}_n$. You will have a chance to work through this in the exercises.

Homomorphisms of polynomial rings

As we consider some examples of homomorphisms of polynomial rings, take a look at Activities 6(d), (e), (f) and (g). Do you see some general principles emerging?

Let $h : R \longrightarrow R'$ be a ring homomorphism. Do you see how to "lift" h to a homomorphism from $R[x]$ to $R'[x]$? Does this have anything to do with Activity 6(d)?

Here is a way to construct a homomorphism $h : R[x] \longrightarrow R$. If p is a polynomial in $R[x]$, define $h(p) = p(0)$, that is, $h(p)$ is the constant term of p. Does this have anything to do with Activity 3 or Activity 6(e)? Could you replace $p(0)$ by $p(a)$ where a is some other fixed element of R?

Finally, could you set up a homomorphism from $R[x]$ to pairs $[a, b]$ of elements of R? How could you make the set of pairs of elements of a ring into a ring? What does this have to with Activity 6(f)? Is there anything analogous that could be done with triples of elements of R?

Embeddings—$\mathcal{Z}$, $\mathcal{Z}_n$ as universal subobjects

The rings $\mathcal{Z}$, $\mathcal{Z}_n$ are especially interesting because, in a sense, one of them can be found inside any ring R with unity. Did you discover this in Activities 4, 5? Given the work we have done on cyclic groups in Section 3.2 (see especially Theorem 3.2), the basic idea is quite simple:

We look at R as an additive group and consider the order of 1 in this group. If the order is finite, n (that is, n is the least number of times 1 has to be added to itself to get 0), we know that the cyclic subgroup generated in R by 1 is isomorphic to $[\mathcal{Z}_n, +_n]$ as groups. (So far, any other non-zero element of R would have done as well.) Using the special properties of 1 we can prove that this subgroup is in fact a subring and that the same isomorphism is also a ring isomorphism. The case where 1 has infinite additive order, and therefore generates a subring isomorphic to $\mathcal{Z}$, is treated similarly.

For completeness, since the terminology for rings is different from that for groups, we introduce it in the next definition.

Definition 5.12 *Given an element r of a ring R and a positive integer k, we denote by kr the result of adding r to itself k times. More generally, for any $k \in \mathcal{Z}$, kr is the k-th power of R in the additive group of R. If there is a positive number n such that $nx = 0$ for all $x \in R$, then the smallest such number is called the* characteristic *of R. If there is no such number, then we say that the characteristic of R is 0.*

Note that if R has an identity 1, then the characterisitic of R is nothing but the order of 1 in the additive group of R, except that when the order is infinite the characteristic is said to be 0.

Theorem 5.5 *Any ring with 1 either has a subring isomorphic to $\mathcal{Z}_n$ for some positive integer n, or has a subring isomorphic to $\mathcal{Z}$.*

Proof. Let R be a ring with 1. First suppose that the characteristic is $n > 0$. Then we already know from Theorem 3.2 (page 98) that the set of additive "powers" of 1,

$$A = \{0, 1 \cdot 1, 2 \cdot 1, 3 \cdot 1, \ldots, (n-1) \cdot 1\}$$

is an additive subgroup isomorphic to $\mathcal{Z}_n$ (as groups). The isomorphism $h : \mathcal{Z}_n \longrightarrow A$ (which in Theorem 3.2 is only discussed implicitly) is given by $h(k) = k \cdot 1$. We now show that this claim also holds for them as rings. First, the relation $n \cdot 1 = 0$, which implied that A is closed under addition, also implies (together with $1 \cdot 1 = 1$) that it is closed under multiplication, hence A is a subring of R. Second, the map h also preserves multiplication, for (using general properties of powers in groups)

$$h(kl) = kl \cdot 1 = (k \cdot 1)(l \cdot 1) = h(k)h(l).$$

The result for characteristic 0 is proved in a similar manner.

$$\square$$

Do you think it is possible for a ring to have subrings isomorphic to two different rings $\mathcal{Z}_n$ and $\mathcal{Z}_k$, or to $\mathcal{Z}_n$ and $\mathcal{Z}$?

The characteristic of an integral domain and a field

Proposition 5.14 *Assume that R is an integral domain with a finite characteristic n. Then n is a prime number.*

Proof. We have $n \cdot 1 = 0$. If n were composite, say $n = mk$ with $1 < k, m < n$, then we would have

$$(m \cdot 1) \cdot (k \cdot 1) = mk \cdot 1 = n \cdot 1 = 0,$$

and since R is an integral domain, $m \cdot 1 = 0$ or $k \cdot 1 = 0$. But this contradicts the minimality of n. Thus n is prime. ▯

Theorem 5.6 *Let F be a field. If F has characteristic p, then F contains a sub-field isomorphic to $\mathcal{Z}_p$. If F has characteristic 0, then F contains a subfield isomorphic to $\mathcal{Q}$.*

Proof. If F has positive characteristic, we know that it must be a prime number p (since every field is an integral domain). We have already seen in Theorem 5.5 that F contains a subring isomorphic to $\mathcal{Z}_p$, and since p is prime, this subring is a field. Assume now that F has characterisic 0. Then Theorem 5.5 says that F contains a subring isomorphic to $\mathcal{Z}$. It is not hard to show that there exists an embedding $f : \mathcal{Q} \longrightarrow F$ satisfying $f(a/b) = ab^{-1}$ for every $a/b \in \mathcal{Q}$. We leave the details to the exercises. ▯

5.3.4 Properties of homomorphisms

Preservation

As was the case with groups (see Section 4.2.4), there are a number of simple properties of ring homomorphisms having to do with preservation of various ring properties. We list these properties in the following proposition and leave their proofs for you to do in the exercises.

Proposition 5.15 *Let $h : R \longrightarrow R'$ be a ring homomorphism, A a subring of R and I' an ideal in R'.*

1. *The image $h(A)$ of A under h is a subring of R'.*

2. *If A is an ideal in R and h is onto, then $h(A)$ is an ideal in R'.*

3. *The inverse image $h^{-1}(I')$ of I' is an ideal in R.*

4. *If R is commutative then so is the subring $h(R)$.*

5. *If R has 1 then $h(1)$ is a multiplicative identity for $h(R)$.*

Proof. Exercise 6.

Ideals and kernels of ring homomorphisms

Since ring homomorphisms are also (additive) group homomorphisms, the notion of kernel of a ring homomorphism is automaticaly defined; that is, the kernel of a homomorphism $h : R \longrightarrow R'$ is the set of all elements of R that go to $0'$, the zero element of R'. The relationship between kernels and ideals is completely analogous to the relationship between kernels and normal subgroups (see Section 4.2.5). Again, we list the relevant properties and leave the proofs as exercises.

Proposition 5.16

1. *The kernel of a ring homomorphism is an ideal.*

2. *A homomorphism is one-to-one if and only if its kernel is $\{0\}$.*

3. *Given a homomorphism $h : R \longrightarrow R'$ and its kernel $I = \ker(h)$, then for $a, b \in R$ we have $a + I = b + I$ if and only if $h(a) = h(b)$. Thus the cosets of I are precisely the inverse images of elements of R'.*

Proof. Exercise 7.

5.3.5 The fundamental homomorphism theorem

The material in this section is very similar to the material in Chapter 4, Section 3. It is more than analogy. In general, this part of the theory of ring homomorphisms is identical to the corresponding theory of group homomorphisms. It is only necessary to check from time to time that the facts remain true when the second operation is considered.

The canonical homomorphism

We begin with the ring analogue of Theorem 4.2, p. 145.

Theorem 5.7 (The canonical homomorphism) *Let I be an ideal in the ring R. Then the map $K : R \longrightarrow R/I$ defined by $K(r) = r + I$ for all $a \in R$ is a ring homomorphism of R onto R/I whose kernel is I.*

Proof. In view of Theorem 4.2, p. 145, it is only necessary to check that the group homomorphism K is also a ring homomorphism. Indeed, in view of the definition of coset multiplication, we have

$$K(rs) = rs + I = (r + I)(s + I) = K(r)K(s)$$

$\square$

We will continue to refer to this homomorphism as the *canonical homomorphism* of R onto R/I (see Definition 4.8, p. 146). It also follows immediately as with groups (see Corollary 4.1, p. 146) that a subset of a ring is an ideal if and only if it is the kernel of some homomomorphism of the ring.

The fundamental theorem

We can also carry over Theorem 4.3, the fundamental homomorphism theorem for groups (see p. 147).

Theorem 5.8 (The fundamental homomorphism theorem.) *Let R and R' be rings, h a homomorphism of R onto R', $I = \ker(h)$. Then R/I is isomorphic to R'.*

Proof. In view of Theorem 4.3, it is only necessary to check that the map $h' : R/I \longrightarrow R'$ defined by $h'(L) = h(r)$, where L is a coset and r is any element of L, preserves multiplication. Recall that the proof of Theorem 4.3 shows that this map is defined (that is, it is independent of the choice of $r \in L$) and is a group homomorphism.

Given L, M in R/I and representatives $r \in L$, $s \in M$, we have $rs \in LM$ so, using the fact that h preserves multiplication, we have,

$$h'(LM) = h(rs) = h(r)h(s) = h'(L)h'(M).$$

$\square$

Homomorphic images of $\mathcal{Z}$, $\mathcal{Z}_n$

The results of the previous paragraph can be used in the same way they were used for groups (see pp. 149ff) to determine all homomorphic images of a given ring by determining its ideals and forming quotients. For the case of $\mathcal{Z}$ and $\mathcal{Z}_n$, however, it is even easier than that because we can use the facts already determined for group homomorphisms of $\mathcal{Z}$ and $\mathcal{Z}_n$. This was dealt with earlier (see Section 5.3.3 pp. 184ff).

Identification of quotient rings

In Activity 6 you had a chance to identify some quotient rings as isomorphic to rings with which you were already familiar. This generally requires two steps. First you have to discover the appropriate familiar ring, and second you must prove the isomorphism. There is no general method for doing the first step. With experience, you will develop the skill of looking at the formation of quotients and guessing what the quotient ring is. Once you have done this, the second step, establishing the isomorphism, is usually not too difficult.

In parts (a), (b) of Activity 6 the point is to determine all of the subrings of $\mathcal{Z}_n, \mathcal{Z}$; note that they are automatically ideals (Theorem 5.3, p. 176). It should be obvious to you what the quotient rings are.

In parts (c), (d) of Activity 6, there is an obvious guess as to what the quotient ring is. For part (c) the guess is the ring $M(R/I)$ of 2×2 matrices with entries in the quotient ring R/I. Does this work? What is a good guess for part (d)?

The last three parts, on polynomial rings, are interesting. For part (e), can you see why a good guess is the base ring R itself? Let's try it and see if it works. We have a ring R, the polynomial ring $R[x]$ and the ideal I of those polynomials whose constant term is 0.

Imagine an arbitrary polynomial $f \in R[x]$ and its coset $f + I$. What other polynomials g are in this coset? Recall that $g \in f + I$ if $f - g \in I$. What does that mean? It means that $f - g$ is a polynomial whose constant term vanishes. What does that say about f, g? About their constant terms? It says that the constant terms are the same. In other words, the coset $f + I$ consists of all those polynomials whose constant term is the same as the constant term of f.

Okay, now here is the thinking part. What does that say about the determination of $f + I$? What is the one single piece of information you need to be able to figure out exactly which polynomials are in this coset? Do you see that the critical piece of information is the value of that constant term? If the coset consists of all the polynomials that have the same constant term, then all we need to know is what that constant term is.

Whatever the term is, at least we know that it is an element of R. This suggests a map. We can take an element a of R and map it to the coset of all polynomials whose constant term is a. We can write this as a formula, since the element a in R can also be considered as a polynomial $\bar{a}$ in $R[x]$, the polynomial with constant term a and all other terms 0. Then we can define

$$h : R \longrightarrow R[x]/I$$

by

$$h(a) = \bar{a} + I.$$

Now there are several things to check: this map is a homomorphism, it is one-to-one, and it is onto. You can do this in the exercises.

There is another way of looking at the same quotient ring and trying to guess "what it really looks like". First we observe that I (the ideal of all polynomials with constant term 0) is actually the set of all polynomials of the form $xg(x)$, the so-called "ideal generated by x". Since I is the zero element in $R[x]/I$, we may think of $R[x]/I$ as what we get from $R[x]$ by "annihilating x". That is, if we start with $R[x]$ and declare x to be 0 (which also forces all its multiples to become 0), then what we end up with is $R[x]/I$. We have thus come full circle. We started by constructing the ring $R[x]$, which is what you get by adjoining an "indeterminate" x to R; and now we proceed to annihilate the very element we added. It is then reasonable to expect that we should end up back where we started, that is back to R. This kind of reasoning (thinking of any quotient ring R/I as what you get from R by annihilating a piece of it), though admittedly somewhat vague, turns out to be extremely powerful in working with quotient rings and in getting to understand them.

At any rate, once you have found the ring R' that you believe is isomorphic to the given R/I, there is a standard way to go about trying to *prove* this conjecture. That is, you try to construct a homomorphism of R onto R' with kernel equal to the given I. If you succeed in this endeavor then you are in luck, since by the homomorphism theorem it follows that $R/I \cong R'$.

The same ideas can be used to identify the quotient rings in Activity 6, parts (f) and (g) except that you need to invent a new way of forming new rings from old. You will have a chance to do this, with some hints, in the exercises.

5.3.6 EXERCISES

1. Determine all the homomorphisms from $\mathcal{Z}$ to any group G.

2. How many ring-homomorphisms are there from $\mathcal{Z}_{20}$ to $\mathcal{Z}_5$? Onto $\mathcal{Z}_5$? From $\mathcal{Z}_5$ to $\mathcal{Z}_{20}$?

3. Check that the map
$$h : R \longrightarrow R[x]/I$$
   defined by $h(a) = \bar{a} + I$ is an isomorphism (see p. 190).

4. Define the ring $\mathcal{Z}[\sqrt{5}]$ the same way we defined $\mathcal{Z}[\sqrt{2}]$ (see p. 166). Prove that these two rings are not isomorphic.

5. Identify the quotient rings in Activity 6, parts (f) and (g) as isomorphic to some familiar rings. (Hint: You will need to invent a new way of forming new rings from old.)

6. Prove Proposition 5.15, p. 187 on invariants of homomorphisms.

7. Prove Proposition 5.16, p. 188 on kernels. (This is analogous to the group situation treated in Section 4.2.5.)

8. We can always define a map $R \rightarrow R'$ by mapping every element of R to $0'$. Show that this map is a ring homomorphism. (It is called the zero or trivial homomorphism.)

9. Prove that a non-trivial homomorphism of a field into a ring must be one-to-one.

10. Prove that $A \oplus B$ is isomorphic to $B \oplus A$.

11. Prove that A and B can be embedded in $A \oplus B$, that is, that there are 1-1 ring homomorphisms from A and from B to $A \oplus B$.

12. Prove that the map $\phi : A \oplus B \rightarrow A$ defined by $\phi((a,b)) = a$ is a homomorphism whose image is A and whose kernel is $\bar{B}$. What is $A \oplus B / \bar{B}$ isomorphic to? (Recall that $\bar{B} = \{(0,b) : b \in B\}$.) Formulate dual results with the roles of A and B exchanged.

13. A homomorphism of a ring R into itself is called an *endomorphism*. The set of all endomorphisms of R is denoted $\text{End}(R)$. Show that it is possible to define operations on $\text{End}(R)$ so that it becomes a ring. How many elements does $\text{End}(\mathcal{Z}_5)$ have? What about $\text{End}(\mathcal{Z}_8)$?

14. Describe all the elements of $\text{End}(\mathcal{Z} \oplus \mathcal{Z})$.

15. Let R be a ring with unity, $a \in R$. Show that a induces an endomorphism λ_a of R via left multiplication. Show that R can be embedded in $\text{End}(R)$. (This is the analog of Cayley's theorem for groups. See Exercise 29, p. 143.)

16. Suppose R is a ring, I an ideal of R. Prove that $M(R)/M(I)$ is isomorphic to $M(R/I)$.

17. Prove that R/J_a is isomorphic to $\mathcal{R}$, where R and J_a are as defined in Exercise 16, p. 180.

18. Prove that for any ring R, $R[x]/\langle x \rangle$ is isomorphic to R.

19. Find familiar rings isomorphic to the following quotient rings (the ideal P is defined in Exercise 17, p. 180).

    (a) $\mathcal{Z}[i]/P$

    (b) $\mathcal{R}[x]/\langle x^2 + 1 \rangle$

    (c) $\mathcal{Z}[x]/\langle x^2 - 2 \rangle$

20. Show that $2\mathcal{Z}/8\mathcal{Z}$ and $\mathcal{Z}_4$ are isomorphic as groups, but not as rings.

21. Find a subring of $M(\mathcal{R})$ which is isomorphic to $\mathcal{C}$. Hint: Look at the matrices of the form $\begin{pmatrix} a & b \\ -b & a \end{pmatrix}$.

22. Find a subring of $M(\mathcal{Z})$ which is isomorphic to $\mathcal{Z}[\sqrt{2}]$. Hint: Look at the matrices of the form $\begin{pmatrix} a & 2b \\ b & a \end{pmatrix}$.

23. Let R be an integral domain and let

$$F_0 = \{[a,b] : a, b \in R \mid b \neq 0\}.$$

Define on F_0 the operations $\oplus$, $\odot$ by

$$[a,b] \oplus [c,d] = [ad + bc, bd]$$

$$[a,b] \odot [c,d] = [ac, bd]$$

where $+$ and juxtaposition refer to the two operations in R.
In addition, define the relation $\sim$ on F_0 by

$$[a,b] \sim [c,d] \text{ when } ad = bc.$$

Develop the *field of quotients of R* by solving the following problems. As you are working on them, ask yourself what these general notions would mean if R were the ring of integers.

(a) Show that $[F_0, \oplus, \odot]$ is an integral domain.

(b) Show that the relation $\sim$ is a congruence relation, that is, all of the following properties hold.

    i. For all $x \in F_0$, $x \sim x$.

    ii. For all $x, y \in F_0$, if $x \sim y$ then $y \sim x$.

    iii. For all $x, y, z \in F_0$, if $x \sim y$ and $y \sim z$, then $x \sim z$.

    iv. For all $w, x, y, z \in F_0$ if $w \sim y$ and $x \sim z$ then $(w \oplus x) \sim y \oplus z$ and $(w \odot x) \sim y \odot z$.

(c) For each element x of F_0 let $[x]$ be the congruence class of x modulo relation. That is,

$$[x] = \{y : y \in F_0 \mid x \sim y\}$$

Show that the set of all congruence classes forms a partition of F_0, that is, the following properties hold.

    i. The union of the congruence classes is F_0.

    ii. No element of F_0 appears in more than one congruence class.

(d) Define two binary operations $\oplus, \odot$ on F so that the following conditions hold.

    i. $(F, \oplus, \odot)$ is a field.

    ii. The original integral domain R is isomorphic to a subring of F.

    iii. If R' is the image of R under this isomorphism, then

$$F = \{x \odot y^{-1} : x, y \in R' \mid y \neq 0\}.$$

(e) Describe F in the case in which R is the ring of integers.

24. Fill in the details of the proof of Theorem 5.6, p. 187.

6

Factorization in Integral Domains

6.1 Divisibility properties of integers and polynomials

6.1.1 ACTIVITIES

1. Write a func dwr that accepts two inputs a, b, checks that they are integers and that b is not zero, and returns the pair [a div b, a mod b].

   Run your func on various number pairs such as $(98, 15)$, $(1024, 256)$, $(78, 176)$.

   Write down a formula that relates a, b, q and r where q is a div b and r is a mod b. In particular, what restriction is there on the possible values for r?

2. For this and the next few activities, you might wish to review Section 5.1.6, p. 161 for the basic facts about polynomials over a general ring. Write **ISETL** funcs to represent the ring $\mathcal{Q}[x]$ of polynomials over $\mathcal{Q}$. In particular, you will need to decide how to represent a polynomial and then to write funcs to represent the degree of a polynomial, polynomial equality and the arithmetic operations on the polynomial ring. Note in particular that your representation should account for the following equalities:

$$x^3 - x + 1 = 0x^4 + x^3 - x + 1 = x^3 + 0x^2 - x + 1.$$

The following few activities give some advice about one way of representing $\mathcal{Q}[x]$ in **ISETL**. However, you are encouraged to try it your own way before looking at this advice.

3. In this activity you will investigate a collection of basic funcs that are used to work in **ISETL** with the ring $\mathcal{Q}[x]$ of polynomials over $\mathcal{Q}$. Before using any of these funcs in **ISETL**, make sure you execute the directive

$$!rational\ on$$

(For details on this directive, see **ISETL** user's manual *Using ISETL 3.0: A Language for Learning Mathematics.*)

A polynomial over $\mathcal{Q}$ will be represented by a tuple of rational numbers

```
f =[f(1),f(2),...f(#f)]
```

and its components will be counted $1, 2, \ldots, \#f$. The corresponding polynomial is
$$f(x) = a_0 + a_1 x + \cdots + a_n x^n$$
and its terms will be counted $0, 1, 2 \ldots, n$. Thus we have the connecting formulae,
$$a_i = f(i+1), \quad i = 0 \ldots, n$$
Note that, if $a_n \neq 0$, the degree of the polynomial is n and the length of the tuple which represents it should be $n + 1$. (Observe, however, that this representation need not be unique in the sense that any component which is 0 could be changed to om or vice-versa without changing the polynomial. The main purpose of the func standard below is to eliminate this non-uniqueness.)

Here are the basic funcs.

```
standard := func(f); $ f is a representation of a
              polynomial if forall a in f | is_defined(a)
              impl a = 0 then
                  return [0];
              end;
              while f(#f) = 0 do f(#f) := om; end;
              for i in [1..#f] do
                  if f(i) = om then f(i) := 0;
              end;
              return f;
          end;
```

```
deg := func(f); $ f is a representation of a polynomial
            if standard(f) /= [0] then return #(standard
            (f)) - 1;
        end;
```

```
mon := func(a,d); $ a is an element of Q and
                  $ d is a non-negative integer
            return standard([0 : i in [0..(d-1)]] + [a]);
        end;
```

Enter each of the above **funcs** in **ISETL** and save them in a file for continued use. For each of the following polynomials, make up at least two different **tuples** to represent them and use these **tuples** to test the **funcs**.

(a)
$$2x^2 + 2$$

(b)
$$\frac{3}{2}x^3 - 3x^2 - 4x + 8$$

(c)
$$x^2 - 4x + 4$$

(d)
$$\frac{3}{2}x^3 - 4x$$

(e)
$$\frac{3}{2}x^5 - 3x^4 - \frac{5}{2}x^3 + 5x^2 - 4x + 8$$

(f)
$$-\frac{2}{5}x^7$$

(In every **func** that you write involving polynomials, you should always apply the **func** **standard** to any input parameter and to any output. Also, before testing two polynomials for equality, be sure to apply **standard** to them.)

After trying your **funcs** on these examples, write a brief statement for each **func** explaining exactly what it does.

4. Write **ISETL** **funcs** padd, psubt, and pmult that will accept two representations of polynomials (in the form of **tuples** of their coefficients) and return the standard representation, respectively, of the sum, difference, or product of these two polynomials.

You may want to read through Section 5.1.6, p. 161 for explanations of these operations.

Remark: In defining these funcs you might wish to make use of the **ISETL** expression x?y which returns the value of x unless it is undefined, in which case it returns the value of y. For example, using expressions of the form a?0 will allow you to treat two polynomials as if they had the same degree.

What can you say about the degree of the sum, difference, and product of two polynomials f, g in terms of the degrees of f and g?

5. In this activity you are to write a func pdiv that will accept two polynomials f, g over Q and, if $g \neq 0$, will return the polynomial obtained by dividing f by g until the remainder is 0 or its degree is less than the degree of g.

   One way to do this is to write a recursive func. You may want to look at the description of recursive funcs in the **ISETL** user's manual *Using ISETL 3.0: A Language for Learning Mathematics*. Following is a recursive formulation of the problem that should help you write the func.

   > Given two polynomials f, g, if $g \neq 0$ the process of division can be described as follows.
   >
   > (a) If f is the zero polynomial, or the degree of f is less than the degree of g, then the answer is the zero polynomial.
   >
   > (b) If the degree of f is greater than or equal to the degree of g, then find a monomial ax^r such that the highest degree term of the product $ax^r g$ is the same as the highest degree term of f. The answer, then, is $ax^r + h$, where h is the result of this process applied to the pair $f - ax^r, g$.

   Explain how this formulation really describes the long division process for polynomials.

   Apply pdiv to the following pairs of polynomials. In each case, do the long division to check if you are correct.

   (a) The pairs f, g where f is the polynomial $\frac{3}{2}x^3 - 3x^2 - 4x + 8$ of Activity 2(b), and g is the polynomial $x^2 - 4x + 4$ of Activity 2(c).

   (b) The pairs f, g where f is the polynomial $x^2 - 4x + 4$ of Activity 2(c), and g is the polynomial $\frac{3}{2}x^3 - 3x^2 - 4x + 8$ of Activity 2(b).

(c) The pairs f, g where f is the polynomial $\frac{3}{2}x^5 - 3x^4 - \frac{5}{2}x^3 + 5x^2 - 4x + 8$ of Activity 2(e), and g is the polynomial $\frac{3}{2}x^3 - 4x$ of Activity 2(d).

(d) The pairs f, g where f is the polynomial $\frac{3}{2}x^5 - 3x^4 - \frac{5}{2}x^3 + 5x^2 - 4x + 8$ of Activity 2(e), and g is the polynomial $2x^2 + 2$ of Activity 2(a).

6. Write a **func** pmod that will do f mod g for polynomials over Q, provided that $g \neq 0$. The resulting polynomial should be the remainder on division of f by g.

   You can do this directly from the definitions, or you can use **pdiv** along with a formula analogous to what you may have found in Activity 1.

   Apply **pmod** to the pairs in Activity 5.

7. You can now revise your **func** dwr so that it takes two inputs, checks if they are both integers and the second one is not 0, or they are both polynomials over Q and the second one is not 0. Your **func** should then return [a div b, a mod b] or [a .pdiv b, a .pmod b] as the case may be.

   Apply your **func** to all the examples in Activities 1 and 5 to be sure that it gives you the same answers you got before.

8. Write a **func** Euc_alg which accepts the same input a,b as dwr (in its revised form). This **func** will be recursive and is specified as follows.

   (a) If the second component in the pair dwr(a,b) is 0 then the answer is b.

   (b) Otherwise the answer is Euc_alg(b, dwr(a,b)(2)).

   You will have to write an auxiliary **func** is_zero to test if an element is 0 (as an integer or a polynomial).

   Apply Euc_alg to all the examples in Activities 1 and 5. Can you give some interpretation to the answers that you get? What does it mean if the answer is 1 (in the case of the integers)?

9. Write out the steps of the process of Euc_alg(a,b) in a sequence of equations, as if you were explaining it to a class of students. Use these equations to try and find x,y (integers or polynomials as the case may be) such that

$$\text{Euc\_alg(a,b)} = \text{xa} + \text{yb}$$

Perform the calculations in some of the examples used in this set of Activities.

Will x,y be the only pair that satisfies this equation?

10. Write an **ISETL** func that will accept an integer greater than 1 and determine if it is prime — that is, not divisible by any other positive integer except itself and one.

   Use your func to find the largest prime that it can find if it runs for no more than 15 minutes.

6.1.2   The integral domains $\mathcal{Z}$, $\mathcal{Q}[x]$

Arithmetic and factoring

There are a number of elementary properties of the ring $\mathcal{Z}$ of the integers with which you are no doubt familiar from your previous experience. Some of these were reviewed and slightly extended in the Activities. We refer to the arithmetic operations of addition, subtraction, multiplication and division (even "long division"). In particular, Activity 1 illustrated the important fact that we can always perform in $\mathcal{Z}$ "division with remainder" (hence the name of the func dwr). Thus if we wish to divide an integer a by a non-zero integer b, we can always find the quotient $q = a$ div b and the remainder $r = a$ mod b. Hopefully you also found in Activity 1 that these integers satisfy the conditions:

$$a = qb + r \text{ and } 0 \le r \le |b|.$$

The familiar elementary properties should also include factoring. You know that there are some integers that are called *primes*. A positive integer p is a prime if it is not 1 and the only positive integers which divide it are p and 1. This familiar definition can be extended to cover negative integers as well, by defining $-p$ to be prime if p is a prime; that is, a non-zero integer p is a prime if it is not 1 or -1 and the only integers which divide it are p, $-p$, 1 and -1. If a prime p divides the product $a \cdot b$ of two integers, then it must divide either a or b. Finally, an integer can always be written as a product of primes. This product is unique up to order and the inclusion of the "trivial" factors 1 and -1.

With some very simple adjustments, every statement in the above paragraphs is meaningful (though not necessarily true) also in the context of an arbitrary commutative ring with identity. There are many such rings in which these statements turn out to be just as true as they are for integers. What we mean by this is not only that the statements are true, but that their proofs are essentially the same as well. It is the purpose of this chapter to investigate rings in which all of these statements can be proved.

There are also some rings for which not all of the statements in the first two paragraphs are true. You will see some examples of these.

In this section, we will concentrate on the ring $\mathcal{Z}$ which gives rise to the above-mentioned factorization issues and the ring $\mathcal{Q}[x]$ of polynomials with rational coefficients, which is a very important generalization. (Actually, anything we do for $\mathcal{Q}[x]$ can be done just as easily for any ring $K[x]$, with K a field.) Then, in the next section we will consider more general situations.

The meaning of unique factorization

Before going on with our plan, however, let us return to the first two paragraphs in this section and eliminate some relatively trivial issues. There is no such thing, for example, as a polynomial $f \in \mathcal{Q}[x]$ which is divisible only by f, $-f$, 1 and -1. Indeed, if a is any non-zero rational number, then we can always write $f = a(\frac{1}{a}f)$, so a divides f. The point is that a is a unit in $\mathcal{Q}[x]$, that is, it has an inverse, and any unit in a ring will divide any element in the ring. This is a trivial exception and we would simply like to rule it out. Do you see the relevance, for the statements in the first two paragraphs, of the fact that in $\mathcal{Z}$ the only units are 1 and -1? What are the units in $\mathcal{Q}[x]$?

Here is how we rule out the "trivial" exceptions. Consider the original statement:

A non-zero integer p is a prime if it is not 1 or -1 and the only integers which divide it are p, $-p$, 1 and -1.

We replace it with the following statement:

A non-zero integer p is a *prime* if it is not a unit and the only integers which divide it are units and integers of the form $u \cdot p$ where u is a unit.

Now, if we replace "integer" by "polynomial in $\mathcal{Q}[x]$", the statement still makes sense and there are some primes in $\mathcal{Q}[x]$. One is $x^2 + 1$ (prove). Can you find any others?

This definition of prime elements can be made in any commutative ring, but it is most appropriate in an integral domain. The question, then, that we will consider in this chapter for $\mathcal{Z}$ and $\mathcal{Q}[x]$ in particular, and integral domains in general, is the truth or falsity of the following three statements:

1. A non-zero element which is not a unit can always be written as a product of primes.

2. This product is unique up to order and the inclusion of factors which are units.

3. If a prime p divides the product $a \cdot b$ of two integers, then it must divide at least one of a or b.

An integral domain in which the first two of these properties are satisfied is said to have *unique factorization*. The last, which is equally important, is a consequence of the first two (Exercise). On the other hand, in some situations, one proves the third and uses it in proving the second (Exercise 5). See, for example, the proof of Theorem 6.4, p. 225.

6.1.3 Arithmetic of polynomials

You can always imagine a polynomial as just a string of its coefficients. Some of the coefficients might be absent which means they are understood to be 0. A polynomial which has exactly one non-zero coefficient is called a *monomial*. The *degree* of a non-zero polynomial f is defined as the highest exponent associated with a non-zero term, and is denoted $\deg(f)$. The zero polynomial is not assigned any degree. All of these facts are concretely implemented by the funcs given in Activity 3.

In order to study factorization of polynomials, the arithmetic operations need to be available. These include the ring operations of addition and multiplication. For integers we assume that they are "second nature" for you, and for polynomials you should have no difficulty adding or multiplying two specific examples. In Chapter 5, Section 1 (p. 161) you can find general descriptions and in Activity 4 you had a chance to implement them as **ISETL** funcs.

There is little to be said about subtraction of polynomials (which you also implemented in Activity 4). You just subtract coefficients of corresponding terms.

Long division of polynomials

With division there *is* something more to be done. There is an algorithm for dividing two polynomials to get a quotient and a remainder, and it is important that it be well understood. One reason for having you write the func pdiv in Activity 5 is to help you solidify your understanding of this process. We review it now.

We begin with an example and try to understand exactly what is going on. A worksheet for dividing the polynomial $2x^6 + \frac{9}{2}x^3 + \frac{3}{8}x + 1$ by $4x^3 - 3x + 1$ might look like the following calculations. Let's try to analyze what is going on.

Notice that there is a sequence of repetitions of the same basic operation. First the polynomial $2x^6 + \frac{9}{2}x^3 + \frac{3}{8}x + 1$ is divided by the polynomial $4x^3 - 3x + 1$. Then the polynomial $\frac{3}{2}x^4 + 4x^3 + \frac{3}{8}x + 1$ is divided by this same $4x^3 - 3x + 1$. Then $4x^3 + \frac{9}{8}x^2 + 1$ is divided by $4x^3 - 3x + 1$.

$$\begin{array}{r} \tfrac{1}{2}x^3 \;+\; \tfrac{3}{8}x \;+\; 1 \end{array}$$

$$x^3 - 3x + 1 \;\big)\; 2x^6 \qquad\qquad +\; \tfrac{9}{2}x^3 \qquad\qquad \tfrac{3}{8}x \;+\; 1$$

$$2x^6 \qquad -\; \tfrac{3}{2}x^4 \;+\; \tfrac{1}{2}x^3$$

$$\tfrac{3}{2}x^4 \;+\; 4x^3 \qquad\qquad +\; \tfrac{3}{8}x \;+\; 1$$

$$\tfrac{3}{2}x^4 \qquad\qquad -\; \tfrac{9}{8}x^2 \;+\; \tfrac{3}{8}x$$

$$4x^3 \;+\; \tfrac{9}{8}x^2 \qquad\qquad +\; 1$$

$$4x^3 \qquad\qquad -\; 3x \;+\; 1$$

$$\tfrac{9}{8}x^2 \;+\; 3x$$

The divisions stop when the polynomial $\tfrac{9}{8}x^2 + 3x$ is the dividend. This is because its degree is smaller than the degree of the divisor $4x^3 - 3x + 1$. The divisions would also stop if it came out "even", that is, if you arrived at a dividend which was the 0 polynomial.

Now look again at the operations that are performed to reduce the dividend. All that happens is that the leading term of the dividend — first $2x^6$, then $\tfrac{3}{2}x^4$, and then $4x^3$ — is divided by the leading term, $4x^3$, of the divisor. This is a monomial division so it always works with rational coefficients (but not, for example, with integers). The answers are, successively, $\tfrac{1}{2}x^3$, $\tfrac{3}{8}x$ and 1. These are put up in the quotient.

Thus the quotient resulting from the division algorithm is the polynomial $\tfrac{1}{2}x^3 + \tfrac{3}{8}x + 1$ obtained from the successive monomial divisions, and the remainder is the polynomial, $\tfrac{9}{8}x^2 + 3x$, which is the first dividend for which no further monomial divisions are possible.

We can summarize these steps for the general case, in which we are given two polynomials f and g in $\mathcal{Q}[x]$ and we wish to find polynomials q and r in $\mathcal{Q}[x]$ such that $f = qg + r$, and $r = 0$ or $\deg(r) < \deg(g)$. Before dealing with the general case, we check all possible special cases that must be handled differently.

1. If $g = 0$, then division cannot take place.

2. If $f = 0$ then no work needs to be done, for we can simply take $q = 0$ and $r = 0$ to get the representation $f = 0 \cdot g + 0$, which is of the required form.

3. If $\deg(f) < \deg(g)$, then the represenation $f = 0 \cdot g + f$ satisfies all the requirements.

4. In all other cases

   (a) Find a monomial m such that multiplying it by the leading term of g gives the leading term of f; that is, mg and f have the same leading term. (This is where the assumption that we are working with polynomials over a *field* becomes absolutely essential. For example, this step cannot always be done in the ring of polynomials over $\mathcal{Z}$.)

   (b) Add the monomial m to the quotient.

   (c) Multiply the monomial by the divisor and subtract the result from the dividend. The result, i.e. the polynomial $f - mg$, is the new dividend. (Can you see why the degree must go down here?)

   (d) Repeat the process from the beginning, but using the new dividend $f - mg$ instead of f.

   (e) Continue this until either the dividend is zero or its degree is less than the degree of the divisor g. (This will terminate the process according to steps 1 or 2.) The current dividend is the remainder r and the accumulated quotient is the required quotient q.

This fairly long description can be streamlined considerably. What is really happening is that, after steps 1, 2 and 3, a monomial m is computed in step 4 and used to create a new dividend. The point is, then, that the quotient of f by g is obtained by adding m to the quotient of $f - mg$ by g. That is all you need to say if you wish to have a recursive formulation of the long division algorithm. This is the description given in Activity 5 and this is what you can use to write a recursive func for long division.

6.1.4 Division with remainder

The process of "long division" for getting a quotient and a remainder, is really quite the same for polynomials as for the integers. In both cases division by 0 is ruled out and the answer is taken to be 0 if the dividend is 0. The procedure consists of successive partial divisions which "reduce" the dividend — in absolute value in the case of integers and in degree in the case of polynomials. The procedure continues until the dividend is "too small" for further division. This last dividend is then the remainder.

In the result of "long division" of a by b (either integers or polynomials), the quotient and the remainder are called, respectively, a div b and a mod b. All these quantities are related by the formula $a = qb + r$, where q is the quotient and r is the remainder. The quantity r is determined by the requirement that it be too small to be divided further by b. In the case of integers, this condition is that $0 \leq r < b$, while in the case of polynomials, the condition is that either $r = 0$ or $\deg(r) < \deg(b)$.

All of this was implemented in Activities 5, 6, and 7.

It is interesting, and important for the definition of division with remainder, that these conditions determine q and r uniquely. This is the content of the following proposition.

Proposition 6.1 *Suppose that we are given two polynomials $a, b \in \mathcal{Q}[x]$ with $b \neq 0$, and that we have the two representations*

$$a = bq_1 + r_1 = bq_2 + r_2$$

where q_1, q_2, r_1, $r_2 \in \mathcal{Q}[x]$, and where $r_1 = 0$ or $\deg(r_1) < \deg(b)$ and $r_2 = 0$ or $\deg(r_2) < \deg(b)$.

Then $q_1 = q_2$ and $r_1 = r_2$.

Proof. Transforming the assumed relation gives

$$bq_1 - bq_2 = r_2 - r_1.$$

We claim that both sides of the equation must be 0 (the zero polynomial). For if they are non-zero, consider their degrees. On the one hand, $\deg(r_2 - r_1) < \deg(b)$ since the degree of a sum or a difference is no higher than the maximal degree of the summands; but on the other hand, $\deg(bq_1 - bq_2) = \deg(b(q_1 - q_2)) \geq \deg(b)$ since for polynomials over a field, the degree of a product is no less than the degrees of each of the factors. This contradiction shows that both sides of the equation must indeed be 0, from which it follows that $r_1 = r_2$ and (since $b \neq 0$) $q_1 = q_2$.

$\square$

This result permits the following definition.

Definition 6.1 *Let a, $b \in \mathcal{Q}[x]$ and suppose that $b \neq 0$. Then $q = a$ div b and $r = a$ mod b are defined to be the (unique) polynomials satisfying the two conditions,*

1. $a = qb + r = (a$ div $b)b + (a$ mod $b)$

2. $r = 0$ or $\deg(r) < \deg(b)$

Almost exactly the same formulation can be given for integers.

Definition 6.2 *Let a, $b \in \mathcal{Z}$ and suppose that $b \neq 0$. Then $q = a$ div b and $r = a$ mod b are defined to be the (unique) integers satisfying the two conditions,*

1. $a = qb + r = (a$ div $b)b + (a$ mod $b)$

2. $0 \leq r < |b|$

The only difference in the two cases is that for polynomials we have used degree and for integers we have used magnitude. The point is that in performing the division, there needs to be a reasonable way of measuring the remainder, to see that it is "smaller" than the dividend. You will see in Section 6.2 that this is all that is needed to make the entire theory of factorization work in an integral domain.

You may have noticed that the statement of condition 2 for polynomials is slightly more complicated than for integers. The reason is that the case $r = 0$ needs to be treated there separately, since the degree of the zero polynomial is not defined.

The case of remainder 0 is important enough to warrant a special terminology.

Definition 6.3 *If a, b, q are elements of an integral domain R such that $a = bq$ then we say that b divides a, and write $b|a$. Alternatively we say that b is a* divisor *of a.*

For example, $7|21$ in $\mathcal{Z}$ and $(x - 1)|(x^2 - 1)$ in $\mathcal{Q}[x]$.

6.1.5   Greatest Common Divisors and the Euclidean algorithm

We have just extended the notion of a divisor from the familiar setting of the integers to an arbitrary integral domain. In this section we would like to do the same for the notion of greatest common divisor (GCD). Looking back at the integers we recall that, for example, the greatest common divisor of 6 and 10 is 2 while that of 3 and 5 is 1. In trying to extend this notion to an arbitrary integral domain some problems immediately present themselves.

The first problem is already in the definition itself, since we don't even know in an arbitrary integral domain what it means for one element to be "greater" than another. What do you think is a reasonable definition of the "greatest common divisor" of two polynomials? (There are at least two good answers to this question.) A second problem is how to compute GCDs in general integral domains, or even just decide if they exist.

Our plan for this section is to start with the definition of a GCD and a few of its elementary properties. Then we present a method for calculating GCDs, which utilizes successive divisions-with-remainders. The method should look pretty familiar to you in light of the func Euc_alg which you wrote in Activity 8. We now give the definition of a GCD for an arbitrary integral domain. Implicit in this definition is the idea that if x divides y then x can in some sense be viewed as being "smaller" than y (or y greater than x).

Definition 6.4 (greatest common divisor.) *Let R be an integral domain and $a, b \in R$. An element $d \in R$ is called a* greatest common divisor *of a, b if d divides both a, b and if any element of R which divides both a, b must also divide d.*

Remark. A given pair of elements a and b may have more than one GCD, though as we shall see later (Prop. 6.8, p. 218) they are all closely related. For this reason we will usually talk about "a GCD" (meaning *any* GCD) or about "GCDs" (meaning *the set* of all GCDs) of a and b.

Proposition 6.2 *Let R be an integral domain. Then*

1. *For every $a \in R$, $a|0$ and $a|a$.*

2. *For all $a, b, c \in R$, if $a|b$ and $b|c$ then $a|c$. In other words, if a divides b then a divides every multiple of b.*

3. *For all $a, b, c \in R$, if $a|b$ and $a|c$ then $a|(b+c)$ and $a|(b-c)$.*

Proof. Exercise 1.

Proposition 6.3 *Let R be either $\mathcal{Z}$ or $\mathcal{Q}[x]$. Then*

1. *For every $a \in R$, a is a GCD of a and 0.*

2. *If $a, b \in R$ and $a|b$, then a is a GCD of a and b.*

3. *Suppose $a, b \in R$, $b \neq 0$, and let $r = a \bmod b$ (i.e. the remainder of a on division by b). Then the GCDs of a and b are the same as the GCDs of b and r.*

Proof.

1. This part clearly follows from the first part of the previous proposition.

2. This part also clearly follows from the definiton of GCD and from the previous proposition.

3. Using division with remainder, we can write $a = qb + r$, where $q = a \operatorname{div} b$. It follows from parts 2 and 3 of the previous proposition, that every common divisor of b and r is also a divisor of a, hence a common divisor of a and b. A similar argument can be given in reverse, using the relation $r = a - qb$. Indeed, it follows from this relation that every common divisor of a and b is also a divisor of r, hence a common divisor of b and r. We conclude that the pairs a, b and b, r have exactly the same common divisors, hence also the same *greatest* common divisors.

$\square$

Remark. Note that items 1 and 2 of the last proposition are true in any integral domain.

The last proposition is the basis of one of the oldest known algorithms, the so-called *Euclidean Algorithm*, which you met in Activity 9. This algorithm is used to calculate a GCD of two given numbers (or polynomials), as well as some other related quantities. The idea is that in order to find a GCD of two elements, we can use part 3 of the last proposition to successively reduce the "magnitude" of these elements until one of them divides the other, whence, by parts 2 and 3, the divisor is a GCD of the original two elements.

We assume then that two elements a and $b \neq 0$ are given, and we perform division-with-remainder of a by b to get the relation

$$a = qb + r.$$

Here either $a, b \in \mathcal{Z}$ or $a, b \in \mathcal{Q}[x]$, and $r = 0$ or r is "smaller" than b, either in magnitude (in $\mathcal{Z}$) or degree (in $\mathcal{Q}[x]$).

We could try to do this again, replacing a by b and b by r, provided that $r \neq 0$. We would then get,

$$b = rq_1 + r_1.$$

If we rename q as q_0 and r as r_0, then we have,

$$
\begin{aligned}
a &= bq_0 + r_0 \\
b &= r_0 q_1 + r_1.
\end{aligned}
$$

If $r_1 \neq 0$, then we could do it yet again. Indeed, as long as the remainder is not 0, we could continue this process. Since each time the remainder becomes "smaller" (in the sense of either the magnitude or the degree), and since the remainder is a positive integer, the process will eventually stop. This can only happen if the remainder is 0. Thus we get the following sequence of relationships.

$$
\begin{aligned}
a &= bq_0 + r_0 \\
b &= r_0 q_1 + r_1 \\
r_0 &= r_1 q_2 + r_2 \\
r_1 &= r_2 q_3 + r_3 \\
&\quad \cdot \quad \cdot \quad \cdot \\
&\quad \cdot \quad \cdot \quad \cdot \\
&\quad \cdot \quad \cdot \quad \cdot \\
r_{n-3} &= r_{n-2} q_{n-1} + r_{n-1} \\
r_{n-2} &= r_{n-1} q_n + r_n \\
r_{n-1} &= r_n q_{n+1}.
\end{aligned}
$$

By part 3 of the last proposition, the following pairs all have the same GCD:

$$(a, b), \ (b, r_0), \ (r_0, r_1), \ (r_1, r_2), \ (r_2, r_3), \dots, (r_{n-1}, r_n)$$

and since $r_n | r_{n-1}$, this GCD is r_n. We conclude that *the last non-zero remainder in the Euclidean Algorithm is a GCD of a and b.*

The Euclidean Algorithm is a practical method for computing GCDs, but it is also one way of proving *existence* of GCDs, which in many theoretical considerations is all we really need. It can also be used for producing a particularly important relation involving the GCD and the two original elements, which you already met in Activity 9. All this is summarized in the following proposition.

Proposition 6.4 *Let R be a ring which is either $\mathbb{Z}$ or $\mathcal{Q}[x]$. Then*

1. *Every pair of elements in R has a greatest commmon divisor in R.*

2. *If $d \in R$ is a GCD of $a, b \in R$, then there exist $x, y \in R$ such that*

$$d = xa + yb.$$

Proof. Part 1, the existence of GCDs, has been in effect already proved. We just collect the pieces. If $b = 0$, then we take a to be a GCD of a and b. Otherwise we apply the Euclidean Algorithm to a and b and we take the GCD to be r_n, the last nonzero remainder. As we have already shown, the Algorithm will always terminate after a finite number of steps and the last nonzero remainder will indeed be a GCD of a, b.

We now prove the second part of the proposition, essentially beginning at the end of the Euclidean Algorithm and "running it backwards". But first we need to dispose of the "trivial" case of $b = 0$. This is easily done since the representation $a = 1 \cdot a + 0 \cdot b$ satisfies all the requirements in this case. We assume, then, that $b \neq 0$ and look at the Euclidean Algorithm for a and b.

We actually can ignore the last equation in the list and use the next to last equation to write, taking $x_n = 1$ and $y_n = -q_n$,

$$d = r_n = x_n r_{n-2} + y_n r_{n-1}.$$

Now, solve the previous equation for r_{n-1} and substitute for r_{n-1} in this equation. Every term has a factor of r_{n-2} or r_{n-3}, so you can collect terms and write,

$$d = r_{n-2} - q_n r_{n-1} = x_{n-1} r_{n-3} + y_{n-1} r_{n-2}.$$

Think about what happened. Starting with an expression for d in terms of r_{n-1} and r_{n-2}, it was possible to solve for r_{n-1} and substitute to convert the equation into an expression for d in terms of r_{n-2} and r_{n-3}. That is, d is equal to some element of R times r_{n-2} plus some element of R times r_{n-3}.

Continuing this, if you solve the previous equation for r_{n-2} and substitute in, you will get an expression for d in terms of r_{n-3} and r_{n-4}. Eventually you get an expression for d in terms of r_0 and r_1. Using the second equation, then, to solve for r_1 you get d in terms of r_0 and b and, finally, using the first equation, you get d in terms of a and b. This proves the second assertion.

$\square$

The last proposition has an alternative formulation in terms of ideals in the ring. This more abstract approach will turn out to be very powerful and will be our main tool in Section 6.2, where most of the material in the present section is abstracted and applied to more general rings.

For now we just mention the easy corollary that in the rings $\mathcal{Z}$ and $\mathcal{Q}[x]$, an ideal generated by two elements can always be generated by one element—their GCD. (The proof is left as an exercise. See Exercises 6 and 8, p. 180 for details concerning these ideals.) A far-reaching extension of this corollary will be formulated and proved in Section 6.2.

We are now ready, in the next section, to consider unique factorization in a general integral domain.

6.1.6 EXERCISES

1. Prove Proposition 6.2, p. 207.

2. Determine the units in $\mathcal{Q}[x]$.

3. Show that $x^2 + 1$ is a prime in $\mathcal{Q}[x]$. Is it a prime in $\mathcal{R}[x]$? In $\mathcal{C}[x]$?

4. Find three primes and five non-primes in $\mathcal{Q}[x]$.

5. Show that in an integral domain, the following condition:

> If a prime p divides the product $a \cdot b$ of two integers, then it must divide at least one of a or b.

implies the condition:

> The decomposition of an element into a product of primes is unique up to order and the inclusion of factors which are units.

6. Show that in an integral domain, the following condition:

> A non-zero element which is not a unit can always be written as a product of primes and this decomposition is unique up to order and the inclusion of factors which are units.

implies the condition:

If a prime p divides the product $a \cdot b$ of two integers, then it must divide at least one of a or b.

7. Prove that if a, b are integers, $b \neq 0$, then the representation $a = qb + r$, $0 \leq r < |b|$ is unique.

8. Discuss the uniqueness of the greatest common divisor in $\mathcal{Z}$.

9. Discuss the uniqueness of the greatest common divisor in $\mathcal{Q}[x]$.

10. Suppose f is a prime polynomial in $F[x]$, where F is a field. Discuss the question of whether f can become non-prime over a subring R of F (assuming that R contains all the coefficients of f).

11. Write an alternative version of the **func** Euc_alg (Activity 8) which uses a **while** loop instead of recursion.

12. Let a, b be elements of an integral domain with $b \neq 0$ and a not a unit. Show that the ideal generated by ab is a proper subset of the ideal generated by b.

13. In an integral domain, show that the two ideals generated by two single elements are the same if and only if the elements are associates.

14. Let R be one of the rings $\mathcal{Z}$ or $\mathcal{Q}[x]$. Show that any ideal in R which is generated by two elements can also be generated by a single element.

15. Let $\mathcal{Q}(\sqrt{2})$ be the smallest subfield of $\mathcal{R}$ containing $\mathcal{Q}$ and $\sqrt{2}$. Does $3 + 4\sqrt{2}$ have an inverse in $\mathcal{Q}(\sqrt{2})$?

6.2 Euclidean domains and unique factorization

6.2.1 ACTIVITIES

1. Write an **ISETL** func is_prime that will test positive integers for being primes.

2. Write **ISETL** func min_div that will accept an integer $n > 1$ and return the smallest divisor (greater than 1) of n. What can you say about the possible values of is_prime(min_div(n)) for various integers n?

3. Write an **ISETL** func decompose that will accept an integer $n > 1$ and return its prime decomposition, that is, a **tuple** of its prime factors, each appearing as many times as it divides n.

   What can you say about the value of the following expression for various values of n: %*decompose(n)?

(Hint: You can use min_div to find the first prime factor p_1, then divide it out and continue the process with n/p_1. In writing the func make sure you use n div p1 rather than n/p1, so that the result is an integer.)

4. Denote by $\mathcal{Z}[i]$ the set of all complex numbers $x + yi$ where $x, y \in \mathcal{Z}$ and $i = \sqrt{-1}$. Write **ISETL** funcs gadd, gneg, gminus and gmult that will implement addition, negation, subtraction, and multiplication of such numbers. An element of $\mathcal{Z}[i]$ is to be represented in **ISETL** by a tuple of two integers. The operations will accept one or two such tuples and return a tuple that represents the sum, negation, difference, or product.

   Use **VISETL** to determine that $\mathcal{Z}[i]$ is an integral domain.

5. Write **ISETL** funcs gconj and gdeg to implement the conjugate $(x - yi)$ and the square of the absolute value $(x^2 + y^2)$ of the element $x + yi \in \mathcal{Z}[i]$. What can you say about the conjugate and about the "degree" of a product of two elements?

6. Investigate the possible values of gdeg(u), where u is a unit in $\mathcal{Z}[i]$. Find all the units in $\mathcal{Z}[i]$.

7. Apply gmult to the set of all pairs of elements in $\mathcal{Z}[i]$ whose "degree" is smaller than or equal to 2. Try to find examples of primes and of non-primes in $\mathcal{Z}[i]$. Check by example the unique factorization properties listed on p. 201. In particular, can you find a decomposition of 7 involving factors other than itself and units? What about 5?

8. Let $\mathcal{Z}\left[\sqrt{-3}\right]$ be the set of all complex numbers of the form $x + \sqrt{3}yi$, $x, y \in \mathcal{Z}$. Use **VISETL** to check that $\mathcal{Z}\left[\sqrt{-3}\right]$ is an integral domain. Repeat all of the tasks in Activities 4, 5, 6, 7 for $\mathcal{Z}\left[\sqrt{-3}\right]$.

9. The purpose of this activity is to write an **ISETL** proc name_eucdom that will construct a func similar to Euc_alg from Section 6.1, Activity 8, p. 199 in any integral domain for which appropriate auxiliary funcs (to be specified below) can be written.

   Your proc is to have the following 7 input parameters: set, add_op, neg_op, mult_op, zero, ident, dwr_op. The first difficulty that arises is with set because we wish to consider infinite integral domains. Hence, the parameter set will be a func which accepts a single input and tests whether it is a representation of some element of the integral domain. You can see how this could be done, for example, with $\mathcal{Z}, \mathcal{Q}[x], \mathcal{Z}[i]$.

   The next three parameters implement addition, negation and multiplication in the integral domain and the two after those are elements of the integral domain. The last parameter is a func which implements division with remainder for the particular integral domain, provided a division algorithm can be found.

Your proc is to construct and name the following: is_R, ad, neg, mul, z, one, minus, dwr and Euc_alg. The first six and the eighth simply rename the input parameters. The seventh, minus, is a func which implements subtraction. The last, Euc_alg, implements an abstract version of the Euclidean algorithm which will work for any integral domain for which the appropriate input parameters to Euc_alg can be found.

Test your proc with the two integral domains $\mathcal{Z}$ and $\mathcal{Q}[x]$, for which most of the required funcs were written in the previous section. Run examples from the activities in Section 6.1 to make sure that you get the same answers as in the original versions.

10. Write a func gdiv that will accept two tuples representing elements of $\mathcal{Z}[i]$ and, if the second does not represent 0, will return a tuple of two *rationals* which represents their quotient. You may find useful the relation $z \cdot \mathrm{gconj}(z) = \mathrm{gdeg}(z)$, from which $1/z$ can be determined, since $\mathrm{gdeg}(z)$ is real.

Also, you will have to execute the directive !rational on before running your func.

Test your func with a number of examples.

11. Revise your func gdiv from Activity 10 so that the pair it returns is not just a pair $[r, s]$ of rationals, but a pair $[x, y]$ of integers such that $|x - r|$ and $|y - s|$ are as small as possible.

Suppose that gdiv is applied to the pair $a, b \in \mathcal{Z}[i]$. What is the largest possible value, in terms of deg(a) and deg(b), that you could get for the following expression?

$$\mathrm{deg}(a \text{ .gminus } (b \text{ .gmult } q))$$

Compare this value to deg(a) and deg(b) in a large number of different examples in order to develop a guess. Then prove that your guess is correct.

12. In light of your experiences with Activity 11 write a func gmod. Write a func gdwr and anything else that is required so that name_eucdom can be applied to $\mathcal{Z}[i]$. Apply the resulting Euc_alg to several examples in $\mathcal{Z}[i]$, taken from Activity 1.

13. Write funcs is_unit and is_assoc to check for units and associates in $\mathcal{Z}$, $\mathcal{Q}[x]$, and $\mathcal{Z}[i]$, and use your funcs to guess answers to the following questions. Try to justify your answers.

(a) If a divides b and b divides c, what can be said about a and c?

(b) If a divides b and a, b are replaced by associates, a', b', does the division property still hold?

(c) If a divides b and b divides a, what can be said about a and b?

(d) What elements can be associates of 1?

6.2.2 Gaussian integers

In our study of factorization so far we have worked with two major examples of integral domains—the integers Z and the polynomials over the rationals $Q[x]$ (or similarly over any other field K). We will now enrich our stock of major examples by considering yet another integral domain—the *Gaussian integers*.

The ring of Gaussian integers consists of the set $Z[i] = \{x+yi : x, y \in Z\}$, with ordinary addition and multiplication of complex numbers. That is, they are all the complex numbers whose real and imaginary parts are both integers. Because the complex numbers form a field, it is easy to deduce that $Z[i]$ is an integral domain. You checked this in Activity 4. Beside the great interest the Gaussian integers have in their own right, they are also an indispensible tool in the application of ring theory to number theory.

In order to calculate with Gaussian integers, it is useful to recall some standard facts about complex numbers in general. In particular, you have programmed in Activity 5 two of the most important quantities defined for any complex number $z = x + yi$: its *absolute value* $|z| = \sqrt{x^2 + y^2}$, which is a non-negative real number, and its *complex conjugate* $\bar{z} = x - yi$. (Actually, you have programmed the "degree" of z, which is the *square* of its absolute value.) In the geometric interpretation of complex numbers (thinking of $z = x + yi$ as the vector from the origin to the point (x, y)) the absolute value of a complex number corresponds to its magnitude and the conjugate to its reflection in the x–axis.

In the following proposition we have collected some of the most important properties of conjugates and absolute values.

Proposition 6.5 *Let z, w be any complex numbers. Then*

1. $\overline{z + w} = \bar{z} + \bar{w}$

2. $\overline{zw} = \bar{z}\bar{w}$

3. $|z| |w| = |z| |w|$

4. $|z + w| \leq |z| + |w|$

5. $z\bar{z} = |z|^2$

Proof. Exercise 1.

We can immediately demonstrate the usefulness of these properties by exihibiting two important applications: Division of complex numbers and finding all units in the Gaussian integers.

To see how to calculate the quotient of two complex numbers z and $w \neq 0$, we employ the familiar "trick" of multiplying numerator and denominator by the conjugate of the denominator (this is permissible because $\overline{w} \neq 0$ as well). The point is that by the last assertion of the proposition, we will then have in the denominator a *real* number:

$$\frac{z}{w} = \frac{z\overline{w}}{w\overline{w}} = \frac{z\overline{w}}{|w|^2}.$$

We can see more explicitly what these calculations mean if we write $z = x + yi$ and $w = u + vi$, so that $\overline{w} = u - vi$ and $|w|^2 = u^2 + v^2$. Thus we have

$$\frac{x + yi}{u + vi} = \frac{x + yi}{u + vi}\frac{u - vi}{u - vi} = \frac{xu + yv}{u^2 + v^2} + i\frac{-xv + yu}{u^2 + v^2}.$$

As a second application of the proposition, we now find all the units in the ring $\mathcal{Z}[i]$ of Gaussian integers. The idea here is that a unit must have absolute value 1. Indeed, if $u \in \mathcal{Z}[i]$ is a unit then it has an inverse $v \in \mathcal{Z}[i]$ such that $uv = 1$. Using the absolute value and the proposition, we get $|uv|^2 = |u|^2|v|^2 = 1$. Since $|u|^2$ and $|v|^2$ are both positive integers, it follows that $|u|^2$ (hence also $|u|$) must be 1. Writing now $u = x + yi$, with x, y integers, we have $x^2 + y^2 = 1$, which can only happen if $x^2 = 1$ and $y^2 = 0$, or $x^2 = 0$ and $y^2 = 1$. Thus the only possible units in $\mathcal{Z}[i]$ are 1, -1, i, and $-i$. (To convince yourself that these are indeed units, list the inverses of these four numbers.)

The question of primes in the Gaussian integers is a bit more involved. Consider, for example, the decomposition $(4 + i)(4 - i) = 16 - (-1) = 17$, so that 17 is not a prime in $\mathcal{Z}[i]$. On the other hand you can prove that $1 + i$ is prime in $\mathcal{Z}[i]$, by imitating our use of absolute values in the discussion of units above. What about 2,3,5? You will have a chance to investigate such questions in the exercises. The inescapable conclusion is that the primes in $\mathcal{Z}[i]$ are very different from the primes in $\mathcal{Z}$.

Nevertheless, we can still ask the question, does unique factorization (in the sense of the properties on p. 201) hold? It is not hard to be fooled on this one. Consider, for example,

$$10 = 2 \cdot 5 = (3 + i)(3 - i).$$

Is this not a case of two different decompositions of 10 into primes? No it is not. You have to decide if $2, 5, 3 + i, 3 - i$ are all primes. They are not! For example, $2 = (1 + i)(1 - i)$.

Here is how we can refine the usual prime decomposition of 10 (in $\mathcal{Z}$) to a prime decomposition in $\mathcal{Z}[i]$.

$$10 = 2 \cdot 5 = (1+i)(1-i)(2-i)(2+i).$$

Can you show that the factors on the right-hand side are all prime?

It is perhaps hard to believe from such examples but, nevertheless, you will see later in this section that unique factorization *does* hold for the Gaussian integers $\mathcal{Z}[i]$.

6.2.3   Can unique factorization fail?

Yes.

If you can have unique factorization in situations as bizarre as a prime decompositon of 10 into a product of four Gaussian integers, then maybe unique factorization is a "fact of nature". Maybe it holds in every integral domain. This is why you were asked to work with $\mathcal{Z}[\sqrt{-3}]$ in Activity 8.

The elements of this integral domain all have the form $x + y\sqrt{-3} = x + \sqrt{3}yi$, where $x, y \in \mathcal{Z}$, and they all satisfy the relation

$$(x + y\sqrt{-3})(x - y\sqrt{-3}) = x^2 + 3y^2$$

(this is just a restatement of $z\bar{z} = |z|^2$ from the last proposition).

Using this relation, it is not too hard to find in $\mathcal{Z}[\sqrt{-3}]$ simple examples of elements which can be expressed as a product of primes in more than one way. Here is one example:

$$4 = 2 \cdot 2 = (1 + \sqrt{-3})(1 - \sqrt{-3})$$

A "neat" way to show that 2, $1+\sqrt{-3}$, and $1-\sqrt{-3}$ are primes in $\mathcal{Z}[\sqrt{-3}]$ is, once again, to utilize absolute values. We leave this as an exercise.

6.2.4   Elementary properties of integral domains

We begin now our discussion of the general theory of factorization in a ring. In the first place, we remark that very little can be done if there is no identity or there are divisors of zero. This is why we restrict the entire discussion to integral domains. It will turn out that unique factorization will hold in some but not all integral domains.

In the next section we will consider a type of integral domain, called Euclidean domain, which forms a very general, but not the most general, class of integral domains for which unique factorization holds. In particular, all the integral domains discussed so far in this chapter—the integers, polynomials over $\mathcal{Q}$ and the Gaussian integers—will turn out to be Euclidean domains. The main idea there will be to consider a generalized form of "division-with-remainder" and show that it implies unique factorization.

Before we embark on this program, however, we need to "put our house in order", by collecting, formalizing and extending the general notions of integral domains—divisors, GCDs, primes, principal ideals—that we have been discussing mainly in the context of polynomials and the integers. This is our task for this section. Some of these definitions have already been given before, but we repeat them here for convenience.

The first definition is intended to capture the idea that multiplying by a unit is in a sense a "trivial" operation, and we would like to view elements differing only by a unit factor as being indistinguishable.

Definition 6.5 *Let R be an integral domain, $a, b \in R$. We say that a and b are* associates, *and write $a \sim b$, if there exists a unit $u \in R$ so that $a = ub$.*

Examples

1. In $\mathcal{Z}$, $a \sim b$ if and only if $a = \pm b$.

2. in $\mathcal{Q}[x]$, two polynomials are associates if and only if one is a multiple of the other by a non-zero "constant" (that is, a rational number).

3. In $\mathcal{Z}[i]$, the numbers $2 + 3i$ and $-3 + 2i$ are associates. Can you find a general rule for associates in $\mathcal{Z}[i]$, like we did for $\mathcal{Z}$ and $\mathcal{Q}[x]$? Can you do this for $\mathcal{Z}[\sqrt{-3}]$?

The following proposition shows that the relation of being associates is an "equivalence relation" on R.

Proposition 6.6 *Let R be an integral domain. Then for all $a, b, c \in R$*

1. *$a \sim a$.*

2. *If $a \sim b$ then $b \sim a$.*

3. *If $a \sim b$ and $b \sim c$ then $a \sim c$.*

Proof. The proof follows easily from the fact that the set of units forms a multiplicative group. We leave the details as exercise.

In light of this proposition we can say, symmetrically, that a and b are associates if they "differ by a unit factor".

The next proposition summarizes the relationship between the two relations "a divides b" and "a and b are associates".

Proposition 6.7 *Let R be an integral domain, $a, a', b, b' \in R$. Then*

1. *If $a|b$, $a \sim a'$, $b \sim b'$ then $a'|b'$.*

2. *$a \sim b$ if and only if $a|b$ and $b|a$.*

Proof. We prove the second assertion, leaving the first as exercise.

If $a \sim b$ then it is clear from the definition that $b|a$. By the previous proposition, we also have $b \sim a$, hence also $a|b$.

Conversely, assume $a|b$ and $b|a$. Then for some $c, d \in R$ we have $b = ca$ and $a = db$. We may assume $a \neq 0$, for otherwise the claim is easily seen to be true. Substituting the value for b in the equation for a, we get $a = dca$, hence $dc = 1$, since the cancellation law always holds in integral domains. Thus c, d are units and $a \sim b$.

$$\square$$

Next we re-state the general definition of prime elements in an integral domain. These are the elements that have no non-trivial decomposition. Note how the definition gets around the obstacle of "trivial" decompositions like $a = u(u^{-1}a)$, in which one factor is a unit and the other an associate of a.

Definition 6.6 (prime element.) *A element p in an integral domain R is said to be* prime *or* irreducible *if it is not a unit and the only elements of R which divide it are either units or associates of p.*

The next proposition states that primality and greatest common divisors are "invariants of associates". It also justifies the convention we adopted in the past, declaring a negative number to be prime if its absolute value is prime. Finally, it establishes that GCDs are unique "up to associates"; for example, 2 and -2 are both GCDs of 4 and 6, and $x + 1, \frac{1}{2}(x+1), 2(x+1)$ are all GCDs of $x^2 + 1$ and $(x+1)^2$. For this reason we will sometimes refer to *the* GCD of a and b, meaning an arbitrary member in the set of GCDs, all of which are associates of each other.

Recall how we defined (p. 206ff) greatest common divisors for an arbitrary integral domain (where the notion of "greatest" by itself may not even have any meaning): d is a GCD of a and b iff d is a common divisor of a and b, and every common divisor of them must also divide d.

Proposition 6.8 *Let R be an integral domain.*

1. *If $p, p' \in R$ are associates and if p is prime then so is p'.*

2. *Let $d \in R$ be a GCD of $a, b \in R$. Then d' is a GCD of a and b if and only if $d' \sim d$.*

Proof. We prove one-half of the second assertion, leaving the rest as exercise. Assume that both d and d' are GCDs of a, b. Then since d is a common divisor and d' is a *greatest* common divisor, we have $d|d'$. Conversely, since d' is a common divisor and d is a greatest common divisor, we also have $d'|d$, hence $d' \sim d$.

$$\square$$

In conclusion of this section on integral domains, we recall the notion of an ideal generated by one or two elements, and show that all the divisibility properties we have been considering can be characterized in terms of these ideals.

Proposition 6.9 *Let R be an integral domain, $a, b \in R$. Then*

1. *The set $\langle a \rangle = \{ra : r \in R\}$ is a minimal ideal of R containing a.*

2. *The set $\langle a, b \rangle = \{ra + sb : r, s \in R\}$ is a minimal ideal of R containing a and b.*

Proof. Exercises 15, 16.

Remark. By "minimal" we mean in the first part that every ideal of R which contains a must also contain $\langle a \rangle$. Similarly, by "minimal" we mean in the second part that every ideal of R which contains a and b must also contain $\langle a, b \rangle$.

Definition 6.7 *Let R be an integral domain, $a, b \in R$. The ideals $\langle a \rangle$ and $\langle a, b \rangle$ defined in the last proposition are called the ideals* generated by a *and by* a, b, *respectively.*

An ideal generated by a single element is called a principal ideal. *That is, to say that an ideal I in R is principal means that there exists an element $a \in I$ such that $\langle a \rangle = I$.*

For example, the ideal $\langle 2 \rangle$ in $\mathcal{Z}$ is the ideal of all the even integers. The ideal $\langle 4, 6 \rangle$ is the set of all integers expressible in the form $4x + 6y$. Can you show that this is actually the same ideal as $\langle 2 \rangle$? Can you find a simple form for the ideal $\langle 6, 9 \rangle$? And the ideal $\langle 3, 5 \rangle$?

The next proposition shows how the properties we have been considering can be characterized in terms of ideals.

Proposition 6.10 *Let R be an integral domain, $a, b, d, p \in R$. Then*

1. *a is a unit if and only if $\langle a \rangle = R$.*

2. *$a|b$ if and only if $\langle b \rangle \subseteq \langle a \rangle$.*

3. *$a \sim b$ if and only if $\langle a \rangle = \langle b \rangle$.*

4. *d is a common divisor of a, b if and only if $\langle a, b \rangle \subseteq \langle d \rangle$.*
   If $\langle a, b \rangle = \langle d \rangle$ then d is a GCD of a, b.

5. *If $\langle p \rangle$ is a prime ideal (see p. 178) then p is a prime element.*

Proof. The proof will consist of a series of hints, giving the main point(s) in each part. You are asked to fill in the details as an exercise.

1. a is a unit if and only if $1 \in \langle a \rangle$ if and only if $\langle a \rangle = R$.

2. If $a|b$ then $b \in \langle a \rangle$. Since $\langle a \rangle$ is an ideal, $br \in \langle a \rangle$ for every $r \in R$. Thus $\langle b \rangle \subseteq \langle a \rangle$.

3. Since $a \sim b$ if and only if $a|b$ and $b|a$, the assertion follows from the previous part.

4. If $\langle a, b \rangle = \langle d \rangle$ then it follows from part 2 of this proposition that d is a common divisor of a, b. It also follows that d has a representation as $xa + yb$ for some $x, y \in R$. Thus every common divisor of a, b must also divide d, hence d is a GCD.

5. We assume that $\langle p \rangle$ is a prime ideal. Recall that this means that if $ab \in \langle p \rangle$ then $a \in \langle p \rangle$ or $b \in \langle p \rangle$. By the previous assertions of this proposition, this translates to the following property of p: If $p|ab$ then $p|a$ or $p|b$. We can now prove that p is prime. Indeed, if we have a decomposition $p = ab$ then, say, $p|a$ and so $a \sim p$. It follows that the only divisors of p are its associates and units, hence p is prime.

$$[]$$

Remark. The proposition left two glaring gaps, where we have stopped short of claiming full equivalence. It is tempting to formulate the missing full versions as questions:

1. Is it true in general that d is a GCD of a, b *if and only if* $\langle a, b \rangle = \langle d \rangle$?

2. Is it true in general that p is a prime element if *and only if* $\langle p \rangle$ is a prime ideal?

These questions turn out to be rather subtle. The answer for a general integral domain is "no", but we will see in the next section that for "Euclidean domains" (which include most of our familiar examples) the answer is positive.

6.2.5 Euclidean domains

Look over your work in Activity 9 and recall the various **ISETL** constructions that were necessary in order to make name_eucdom work. With one exception, everything is straightforward and could be done for just about any integral domain (including the device of replacing a set by a test in case of an infinite set).

The only thing that has to be thought through separately is the **func** dwr. The basic idea is that there should be a division algorithm that lets you perform a division to get a quotient and a remainder. The remainder, if not zero, must be, in some measure, smaller than the divisor. Both the algorithm and the measure are different in each example and there are some integral domains for which this is impossible. For integers, the algorithm is predefined in **ISETL** as div, and the measure is absolute value. For polynomials over Q the algorithm is long division of polynomials and the measure is degree. For Gaussian integers, the algorithm is introduced in Activities 10, 11, and 12 where use is made of the measure, $x^2 + y^2$ (the square of the absolute value in the sense of complex numbers).

With these thoughts in mind, we engage in a typical act of mathematical abstraction. First we decide that the essential properties that an integral domain needs in order for the Euclidean algorithm to work consists of the existence of a measure of some sort (which we will call degree) and a division algorithm that produces a quotient and a remainder which, if not zero, is smaller in "degree" than the divisor. Then we construct a subclass of integral domains for which we *postulate* the existence of these two tools. These will be our Euclidean domains. Then the work branches in two directions. One is to verify that various familiar examples (Z, $Q[x]$, $Z[i]$) are Euclidean domains, and the other is to show that the properties we are after, such as unique factorization, hold for any Euclidean domain, and therefore for the domains that we have been considering.

We embark now on this program and we begin by defining the class of integral domains in which we will study unique factorization. Essentially, the defining property of this class is nothing more than the assertion that "division with remainder" is possible.

Definition 6.8 *An integral domain R is called a Euclidean domain if for every non-zero element $a \in R$ there is assigned a non-negative integer, called* the degree of a *and denoted* $\deg(a)$, *such that the following conditions are satisfied:*

1. $\deg(a) \leq \deg(ab)$, *for all nonzero* $a, b \in R$.

2. *If* $a, b \in R$ *and* $b \neq 0$, *then there exist* $q, r \in R$ *such that*

$$a = bq + r \text{ and } r = 0 \text{ or } \deg(r) < \deg(b)$$

Examples of Euclidean domains

The work of Section 6.1 establishes that Z and $Q[x]$ are Euclidean domains with absolute value and degree, respectively, serving as the degree function. We will see at the end of this section that $Z[\sqrt{-3}]$ is *not* a Euclidean domain. We turn now to showing that $Z[i]$ is a Euclidean domain.

Proposition 6.11 $Z[i]$ *is a Euclidean domain.*

Proof. We define the degree function by $\deg(x + yi) = |x + yi|^2 = x^2 + y^2$ for all $x + yi \in Z[i]$. Since it is true for complex numbers in general that $|zw| = |z||w|$, we also have $\deg(ab) = \deg(a)\deg(b) \geq \deg(a)$. Thus the first condition of Definition 6.8 is satisfied.

The second condition of Definition 6.8 requires the construction of a division algorithm. This was actually done in Activity 12 and we need only check the details (see the remark following the proof for the simple idea behind this algorithm). We start with any a and $b \neq 0$ in $\mathcal{Z}[i]$. We will write

$$a = qb + r$$

with q and r defined as follows. First we divide a by b as two complex numbers, that is, we write $\frac{a}{b} = x + yi$, with $x, y \in \mathcal{Q}$. (In Section 6.2.2 we actually developed explicit expressions for x and y, but we shall not need them here). Next we define $q = k + ni \in \mathcal{Z}[i]$ where k is an integer whose distance from x is minimal and n is an integer whose distance from y is minimal. Finally, we define $r = a - bq$. This definition guarantees that $a = bq + r$ and that $r \in \mathcal{Z}[i]$ (since $a, b, q \in \mathcal{Z}[i]$). It only remains to show that $\deg(r) < \deg(b)$. (See the second remark following this proof for a "geometrical" way of looking at the calculation that follows.)

To prove the required inequality, we write $r = a - bq = b(\frac{a}{b} - q)$ and compute:

$$\deg(r) = \deg(b(\frac{a}{b} - q)) = \deg(b)\deg(\frac{a}{b} - q).$$

It is enough, therefore, to show that $\deg(\frac{a}{b} - q) < 1$. Indeed, recall that k and n were chosen as the integers nearest to x and y respectively. Since a rational number is never farther than $\frac{1}{2}$ from some integer, this means that $|x - k| \leq \frac{1}{2}$ and $|y - n| \leq \frac{1}{2}$. Thus

$$\frac{a}{b} - q = (x + yi) - (k - ni) = (x - k) + (y - n)i$$

and

$$\deg(\frac{a}{b} - q) = (x - k)^2 + (y - n)^2 < \frac{1}{2} + \frac{1}{2} = 1.$$

$$\square$$

Remark. The idea behind the division algorithm is simple. Given a and $b \neq 0$ in $\mathcal{Z}[i]$, we can actually carry out *full* division (i.e. without remainder) within the field of the complex numbers, that is, we can express the quotient a/b in the form $x + yi$. Trouble is, the coefficients x and y will not in general be *integers*, so the resulting number $x + yi$ fails to be in $\mathcal{Z}[i]$. Our solution to the problem is to *approximate* $x + yi$ by a Gaussain integer q (taking for its coefficients the nearest integers to x and y) and push the remaining "error" to the remainder r. Note that the success of this scheme depends entirely on whether the degree of the resulting remainder will turn out to be lower than that of the divisor.

Remark. The intuition behind the calculation in the proof that $\deg(r) < \deg(b)$ comes from viewing complex numbers as vectors in the plane and their absolute value (hence also our "degree") as measuring their magnitude. In the same vein, the degree of $u - v$ measures the distance between u and v. Thus the required inequality essentially follows from the fact that we can write $r = b(\frac{a}{b} - q)$, where the second factor is "small" since q was chosen to be "near" $\frac{a}{b}$.

To see how fragile these calculations are, we ask you in the exercises to investigate whether the same method still works to show that $\mathbb{Z}[\sqrt{-3}]$ is a Euclidean domain.

6.2.6   Unique factorization in Euclidean domains

We are now ready to address the main goal of this chapter so far—the proof that Euclidean domains have unique factorization. We first give a characterization of units, divisibility and associates in terms of the degree function in a Euclidean domain.

Proposition 6.12 *Let R be a Euclidean domain. Then the following statements are true.*

1. *$u \in R$ is a unit iff $\deg(u) = \deg(1)$.*

2. *If $a|b$ then $\deg(a) \leq \deg(b)$ with equality holding if and only if $a \sim b$.*

Proof. We first prove the second part. The claim that $a|b$ implies $\deg(a) \leq \deg(b)$ is just a rewording of the defining property $\deg(x) \leq \deg(xy)$. Now to the case of equality. If $a \sim b$ then $a|b$ and $b|a$ and the equality of their degrees follows from what we just proved. Conversely, assume that $a|b$ and $\deg(a) = \deg(b)$. To show $a \sim b$, we invoke division-with-remainder to show that $b|a$. Indeed, if b doesn't divide a, then we can write $a = qb + r$ with $\deg(r) < \deg(b) = \deg(a)$. On the other hand, since $a|b$ and $r = a - qb$, we see that $a|r$, hence $\deg(a) \leq \deg(r)$. This contradiction shows that we must have $b|a$ and, since $a|b$ is given, $a \sim b$.

Finally, the first assertion follows from the second, since $1|u$ and u is a unit if and only if $u \sim 1$.

$$\square$$

We now state and prove some of the most important theorems of this chapter. For the next theorem, you may want to review the definition of a principal ideal (Definition 6.7, p. 219.)

Theorem 6.1 *In a Euclidean domain R every ideal is principal.*

Proof. Let I be an ideal in R. Since the thorem is trivially true for $I = \{0\}$, we may assume that I contains a nonzero element a.

Let $m \in I$ be a nonzero element with minimal degree; that is, no nonzero element of I has a smaller degree than m. We can always find such an element since the degrees of the elements of I are non-negative integers.

We claim that $I = \langle m \rangle$. Indeed, the inclusion $\langle m \rangle \subseteq I$ is clear since $m \in I$. To show the opposite inclusion, we take any element $x \in I$ and we need to show that $x \in \langle m \rangle$, that is, $m|x$. It should come as no surprise at this stage that we show this by doing division-with-remainder and showing that the remainder must be 0. Indeed, let $x = qm + r$, where $q, r \in R$ and $r = 0$ or $\deg(r) < \deg(m)$. But $r = x - qm$ is in I (since $x, m \in I$ and I is an ideal), hence r cannot have a smaller degree than m (since we chose $\deg(m)$ to be minimal in I). We conclude that $r = 0$, $m|x$ and $I = \langle m \rangle$.

$\square$

Remark. In Proposition 6.4, p. 209, we used the Euclidean Algorithm to show that in the domains $\mathbb{Z}$ and $\mathbb{Q}[x]$, every pair of elements a, b has a GCD d, which can be expressed as $d = xa + yb$. We are now going to show that the same is true in any Euclidean domain. One way to achieve this would be by showing that the Euclidean Algorithm is still valid in an arbitrary Euclidean domain. Instead, we choose a more abstract approach, deducing the required properties of GCDs almost immediately from the above "principal ideal theorem". The simplicity of the resulting argument may give you a glimpse of the great power of the abstract approach in general, and of ideal theory in particular. The main theorem of Section 6.3 (Theorem 6.6, p. 235) will give you another glimpse.

Theorem 6.2 *In a Euclidean domain R, every pair of elements $a, b \in R$ has a greatest common divisor d, which can be represented as $xa + yb$ for some $x, y \in R$.*

Proof. Given $a, b \in R$, we consider the ideal $\langle a, b \rangle$ they generate. By the last theorem, there exists $d \in R$ which generates this ideal:

$$\langle a, b \rangle = \langle d \rangle.$$

As we have seen in Proposition 6.10, p. 219, this implies both that d is a GCD of a, b and that it has a representation in the form $xa + yb$.

$\square$

This theorem partially answers one of the questions we left open in a remark on p. 220. It shows that in a Euclidean domain the conditions "d is a GCD of a, b" and "$\langle a, b \rangle = \langle d \rangle$" are indeed equivalent.

As a corollary of the last theorem we can now prove that a crucial property of prime elements—one you might be familiar with from the integers—is true in any Euclidean domain.

Theorem 6.3 *In a Euclidean domain, whenever a prime element divides the product of two elements, it must divide one of the elements. (In symbols: if $p|ab$ then $p|a$ or $p|b$.)*

Proof. Let p, a, b be elements in a Euclidean domain R, and assume that p is prime. We need to show that if $p|ab$ then $p|a$ or $p|b$. We assume that p does not divide a and deduce that $p|b$. Indeed, if p does not divide a, then since p is prime, it follows that 1 is a GCD of p and a. By the last theorem, we have a representation $1 = xp + ya$, and multiplying through by b we get

$$b = xpb + yab.$$

Since $p|ab$, it follows that $p|xpb + yab$, hence $p|b$.

□

Remark. It is easy to extend the last proposition to products of any finite number of elements. That is, if a prime p in a Euclidean domain divides any finite product, then it must divide one of the factors (Exercise 5).

We are now ready to prove our main theorem on unique factorization for Euclidean domains, but first we give a precise definition of "unique factorization".

Definition 6.9 *An integral domain R is said to be a* unique factorization domain *if it satisfies the following properties:*

1. *Every non-zero element of R which is not a unit can be written as a product of (one or more) primes.*

2. *The representation of an element of R as a product of primes is unique up to order and associates; that is, if an element a has two such representations*

$$a = p_1 p_2 \cdots p_n = q_1 q_2 \cdots q_m$$

where $p_1, p_2, \ldots, p_n, q_1, q_2, \ldots, q_m$ are all primes, then $n = m$ and $q_1, q_2, \ldots, q_m$ can be re-ordered so that

$$p_1 \sim q_1, \; p_2 \sim q_2, \ldots, p_n \sim q_n$$

Theorem 6.4 *Any Euclidean domain is a unique factorization domain.*

Proof of Existence. This part of the proof imitates the familiar process of factoring an integer into its prime factors, which you implemented in Activity 3. Our task here is to show that the same process is possible in any Euclidean domain.

Let a be nonzero element of R which is not a unit. If a is prime, then $a = a$ is a representation as required and we are finished. Otherwise, we claim that a must have a prime divisor. Indeed, let p_1 be any divisor of a having *minimal degree*. Then p_1 must be prime since any proper divisor of p_1 would be a divisor of a with a degree lower than that of p_1. We thus have a decomposition $a = p_1 a_1$, where p_1 is prime and $\deg(a_1) < \deg(a)$. The last inequality follows from Proposition 6.12, p. 223, since a_1 divides, but is not an associate of a.

We repeat the same process on a_1. That is, if a_1 is prime then $a = p_1 a_1$ is a representation as required and we are finished. Otherwise, the same argument yields a decomposition $a_1 = p_2 a_2$, or $a = p_1 p_2 a_2$, where p_2 is prime and $\deg(a_2) < \deg(a_1) < \deg(a)$.

We repeat the process on a_2, a_3, and so on as long as we can, yielding in the k-th step a decomposition $a = p_1 p_2 \cdots p_k a_k$, where the degrees of the a_i's satisfy

$$\deg(a) > \deg(a_1) > \deg(a_2) > \cdots > \deg(a_k).$$

Such a strictly decreasing sequence of non-negative integers must terminate after a finite number of steps, which can only happen if the process eventually produces a prime a_i, hence a prime decomposition for a.

Proof of Uniqueness. It suffices to show that if we have two non-empty sets of primes P and P' whose product, together with products of units, is the same, then it is possible to set up a one-to-one correspondence between the two sets so that corresponding elements are associates. Indeed, if we pick any $p \in P$, then p divides the product of the primes in P, and hence it must also divide the product of the primes in P'. By Theorem 6.3, p. 224 (cf. also the remark following the proof), it follows that p divides some $p' \in P'$. Since both are primes, it follows that they must be associates. We now remove this pair from their respective sets and get two smaller sets $P - \{p\}$ and $P' - \{p'\}$ whose product (possibly together with some units) is still the same (why?). If these new sets are still non-empty, we can repeat the same process once again and remove another pair of associate primes. This process continues until we reach the last prime in P. We leave it as an exercise to show that at this point P' must also consist of a single prime and that these two primes are associates. This completes the proof of the theorem.

$$\square$$

The following corollary summarizes what this theorem tells us about the examples we have been considering in this chapter.

Corollary 6.1

1. *The rings $\mathcal{Z}$, $\mathcal{Q}[x]$ and $\mathcal{Z}[i]$ are unique factorization domains.*

2. *The ring $\mathcal{Z}[\sqrt{-3}]$ is not a Euclidean domain.*

Proof. The first assertion follows from the last theorem, since we have seen that $\mathcal{Z}$ and $\mathcal{Q}[x]$ are Euclidean domains. The second assertion follows from the last theorem since we have seen p. 216 that $\mathcal{Z}[\sqrt{-3}]$ is not a unique factorization domain.

$\square$

6.2.7 EXERCISES

1. Prove Proposition 6.5 on p. 214.

2. Prove the first part of Proposition 6.7, p. 217.

3. Complete the proof of Proposition 6.8, p. 218

4. Fill in the details of the proof of each part of Proposition 6.10, p.219.

5. Show that if a prime p in a Euclidean domain divides a product of n elements, then it must divide at least one of them.

6. Find all primes $x + yi \in \mathcal{Z}[i]$ for which $x^2 + y^2 \leq 25$.

7. Determine whether 2,3,5 are primes in $\mathcal{Z}[i]$.

8. Show that the only units in $\mathcal{Z}[\sqrt{-3}]$ are $1, -1$.

9. Show that $\alpha = 1 + \sqrt{-3}$ is prime in $\mathcal{Z}[\sqrt{-3}]$. (Hint: Assume that we have a decomposition $\alpha = \beta\gamma$, for some $\beta, \gamma \in \mathcal{Z}[\sqrt{-3}]$. Use absolute values to show that one of them must be 1 and the other 4.)

10. Show that 2 is prime in $\mathcal{Z}[\sqrt{-3}]$.

11. Show that the ideal generated by a set exists and is unique.

12. Prove that the definition of degree in the proof of Proposition 6.11, p. 221 satisfies the first condition of Definition 6.8, p. 221.

13. What happens if you try to imitate the proof of Proposition 6.11, p. 221, in an attempt to show that $\mathcal{Z}[\sqrt{-3}]$ is a Euclidean domain?

14. Complete the proof of the uniqueness of the prime decomposition in Theorem 6.4, p. 225.

15. Prove part 1 of Proposition 6.9, p. 219 characterizing the ideal generated by a single element.

16. Prove part 2 of Proposition 6.9, p. 219 characterizing the ideal generated by two elements.

17. Show that a union of an increasing sequence of ideals in a ring is again an ideal.

18. Show that in a Euclidean domain, there cannot be an infinite strictly increasing sequence of ideals. Use this result to give an alternative proof that Euclidean domains have prime decomposition.

19. Show that if p is a prime then any $x^n - p$ is irreducible in $Q[x]$.

20. Is $x^2 + x + 4$ a prime in $Z_{11}[x]$? If not, write it as a product of prime factors.

21. Is $x^3 + 6$ a prime in $Z_7[x]$? If not, write it as a product of prime factors.

22. Determine the irreducible polynomials over Z_p, p a prime, of the form $x^2 + ax + b$.

23. Determine the irreducible quadratic polynomials over Z_p, p a prime.

24. Prove Proposition 6.6, p. 217, that "associate" defines an equivalence relation.

25. Determine the units of $Z[\sqrt{d}]$ where d is an integer less than -1 and is not divisible by the square of any prime.

26. Show that if I is a proper ideal in $Z[i]$, then $Z[i]/I$ is finite.

27. Show that the set of all polynomials whose coefficients are even integers is a prime ideal in $Z[x]$.

28. Show that the set $\{a + b\sqrt{-5} \mid a, b \in Z, a - b \text{ is even}\}$ is a maximal ideal in $Z[\sqrt{-5}]$.

29. Show that 2 is not an element of the ideal in $Z[\sqrt{-5}]$ generated by $4 + 6\sqrt{-5}$.

30. Show that $(3 + 2\sqrt{2})^5$ is a unit in $Z[\sqrt{2}]$. Can you generalize this?

31. Is the ideal in $Z[\sqrt{2}]$ generated by $\sqrt{2}$ maximal?

32. Is the ideal in $Z[\sqrt{3}]$ generated by $\sqrt{2}$ maximal?

33. Is the ideal in $Z[i]$ generated by 3 maximal?

34. Is the ideal in $Z[i]$ generated by 6 maximal?

35. Must a subdomain of a Euclidean domain be Euclidean domain? Justify your answer.

6.3   The ring of polynomials over a field

The abstract analyses in this book can be applied to many topics that are important in high school mathematics and beyond. In this section we would like to develop an abstract setting for three major topics, which might be familiar to you from high school: factorization, zeros (or roots) of polynomials, and the complex numbers.

6.3.1   Unique factorization in $F[x]$

If F is a field, then we have seen that the ring $F[x]$ of polynomials with coefficients in F is an integral domain (Proposition 5.8, p. 163). The main example we have considered is the case $F = Q$, the field of rational numbers. You saw in the first two sections of this chapter that in this case, $Q[x]$ has unique factorization. In fact, this remains true if Q is replaced by an arbitrary field F. Moreover, the proof in the general case is the same as the proof for $Q[x]$.

Let's review that proof. All of the essential ideas are in Section 6.1 and the formal argument is in Section 6.2. The culmination of the proof was the application of Theorem 6.4, p. 225, which reduces the problem to showing that $Q[x]$ is a Euclidean domain. Repeating the same kind of reduction here, it is only necessary for us to determine that $F[x]$ is a Euclidean domain.

According to Definition 6.8, p. 221, it is necessary to define a degree function which satisfies the following two conditions.

1. $\deg(f) \leq \deg(fh)$, for all nonzero f, h in $F[x]$.

2. If $f, h \in F[x]$ and $h \neq 0$, then there exist $q, r \in F[x]$ such that

$$f = hq + r \text{ and either } r = 0 \text{ or } \deg(r) < \deg(h).$$

As with $Q[x]$, we use the degree of the polynomial which, as you may recall, has been defined as the highest exponent appearing in a nonzero monomial. (The degree of the zero polynomial is undefined).

The two properties required for $F[x]$ to be a Euclidean domain are established in the following proposition.

Proposition 6.13 *Let F be a field. Then the polynomial ring $F[x]$ is a Euclidean domain and hence also a unique factorization domain.*

Proof. We show that the degree function for polynomials satisfies the two conditions of Definition 6.8. To show the first condition, we establish an even stronger condition: the degree of a product of two nonzero polynomials is equal to the sum of the two degrees. In fact, let f, g be nonzero polynomials in $F[x]$. If the leading term of f is $a_n x^n$ and the leading term of g is $b_m x^m$, then the leading term of fg is $a_n b_m x^n x^m = a_n b_m x^{n+m}$. (Can you see what property of fields is used in the last derivation?)

The second condition of the definition of Euclidean domain is that division with remainder is possible. Here, we can repeat the description given on p. 203 which works for a ring of polynomials over any field: Let $f, h \in F[x]$ with $h \neq 0$. If $f = 0$ then we can take $q = r = 0$ and write

$$f = hq + r.$$

If the degree of f is already less than the degree of h then we get the same relation with $q = 0$ and $r = f$. Finally, if the degree of f is strictly larger than that of h we perform long division as described on p. 202, building up the quotient by adding monomials that successively reduce the degree of the dividend until it is 0 or less than the degree of g. This gives a remainder as desired.

$\Box$

Remark. The above proposition applies in particular to show that the ring $\mathcal{R}[x]$ of polynomials with real coefficients has unique factorization. Similarly, the proposition applies to the field $\mathcal{Z}_p$, p a prime, and so we may conclude that $\mathcal{Z}_p[x]$ has unique factorization.

Notice, however, that since $\mathcal{Z}$ is not a field, we cannot conclude that $\mathcal{Z}[x]$ is Euclidean. Therefore, this argument does not show that $\mathcal{Z}[x]$ has unique factorization. It is possible to give another (more intricate) argument which proves that $\mathcal{Z}[x]$ does have unique factorization after all. Notice, too, that the question as to whether $\mathcal{Z}[x]$ is a Euclidean domain is still open, since the failure of one proof doesn't make a theorem false. However, it can be proved that indeed $\mathcal{Z}[x]$ is *not* Euclidean. One such proof consists of showing that the GCD of x and 2 (which is 1) cannot be expressed in the form $t \cdot 2 + r \cdot x$, and then invoking Theorem 6.2, p. 224. We leave the details for the exercises.

6.3.2   Roots of polynomials

The fact that in $F[x]$ division with remainder is always possible leads to a number of results that, for the case $F = \mathcal{R}$, the field of real numbers, form part of high school mathematics. We begin with some simple definitions, concerning the substitution of a field element in a polynomial.

Definition 6.10 *Assume we have a polynomial $f \in F[x]$ given by*

$$f(x) = a_n x^n + a_{n-1} x^{n-1} + \cdots + a_1 x + a_0$$

and $c \in F$. Then we write

$$f(c) = a_n c^n + a_{n-1} c^{n-1} + \cdots + a_1 c + a_0$$

(an element of F), and we say that $f(c)$ is obtained by substituting c for x in f.

We say that c is a root *or* zero *of f if $f(c) = 0$.*

Remark. We shall also be using the following terminology. A polynomial in $F[x]$ is referred to as a polynomial *over* F. A prime polynomial is also called an *irreducible* polynomial (See Definition 6.6, p. 218.)

Proposition 6.14 *Assume F is a field, $f \in F[x]$, $c \in F$. Then:*

1. The remainder on dividing f by $x - c$ is $f(c)$.

2. *c is a root of f if and only if $x - c$ divides f.*

3. *The number of distinct roots of a nonzero polynomial in $F[x]$ is at most equal to its degree.*

Proof.

1. We can divide $f(x)$ by $x - c$, obtaining

$$f = (x - c)q + r$$

where, since the degree of $x - c$ is 1, either $r = 0$ or r is a polynomial of degree 0. In either case, we have $r \in F$. Substituting c in the polynomials on both sides of the equation, we get

$$f(c) = 0 \cdot q(c) + r = r.$$

(Why is it permissible to substitute c in an equation? This question will be discussed in the next section.)

2. By definition, $x - a$ divides f if and only if $r = 0$. Since $r = f(a)$ this statement follows from the previous statement.

3. If $a_1, a_2, \ldots, a_n$ are the distinct roots of f we already know that each $x - a_i$ divides f. We will show that, moreover, their *product* also divides f. Since $x - a_1 | f$, we can write $f = (x - a_1)g$, for some polynomial $g \in F[x]$. Now a_2 is a root of f but not of $x - a_1$ (since $a_1 \neq a_2$). Substituting a_2 into the above equation we conclude that a_2 must be a root of g. Thus $g = (x - a_2)h$ for some $h \in F[x]$ and $f = (x - a_1)(x - a_2)h$. In a similar way we can now show that a_3 is a root of h and so forth. Eventually we get a polynomial $q \in F[x]$ such that

$$f = (x - a_1)(x - a_2) \cdots (x - a_n)q.$$

It follows that n cannot be larger than the degree of f.

$$\square$$

Remarks.

1. It is possible to give another proof of the last part of the proposition, using the fact that $F[x]$ has unique factorization. For this you need to show that each $(x - a_i)$ is irreducible and that no two are associates. Then you can use unique factorization to conclude that $(x - a_1)(x - a_2) \cdots (x - a_n)$ divides f. We leave the details as exercise.

2. A simple consequence of the proposition is that if a polynomal $f \in F[x]$ of degree higher than 1 has a root $a \in F$ then, since $(x - a)|f$, f is reducible. The converse is not generally true, except for very small

degrees. Can you see that a reducible polynomial over F of degree 2 or 3 must have a root in F? In particular, since $x^2 + 1$ has no root in $\mathcal{R}$, $x^2 + 1$ is irreducible in $\mathcal{R}[x]$. That such implications are not true in general can be seen from the example of the polynomial $(x^2 + 1)^2$ which is clearly reducible over $\mathcal{R}$, but has no roots in $\mathcal{R}$ (why?)

6.3.3   The evaluation homomorphism

In the proof of the last proposition we made use several times of the device of substituting an element in a polynomial equation. The fact that this is permissible is not immediately clear from the definition of substitution, so we now turn to discuss it. What we need to know is, that if we have an equality between two expressions, each made up of sums and products of polynomials over F, then substituting an element of F in the two expressions will preserve the equality. Can you think of a more precise way of formulating such a statement? Does the wording "an operation that preserves sums and products" remind you of something? Can you see the connection of all this to the following proposition?

Proposition 6.15 *Let F be a field, a an element of F, and let us define the map $\phi : F[x] \to F$ by $\phi(f) = f(a)$ for every $f \in F[x]$. Then*

1. *ϕ is a (ring) homomorphism of $F[x]$ onto F.*

2. *$\phi(b) = b$ for all $b \in F$.*

3. *The image of the polynomial x under ϕ is a.*

Proof. To show that ϕ is a homomorphism, we need to show that

$$\phi(f + g) = \phi(f) + \phi(g) \text{ and } \phi(f \cdot g) = \phi(f) \cdot \phi(g).$$

That is, we need to show that

$$(f + g)(a) = f(a) + f(b) \text{ and } (f \cdot g)(a) = f(a) \cdot g(a).$$

The first assertion follows easily from the definitions of polynomial addition and of substitution. The second asssertion also follows from the definitions, but is more messy to prove due to the relatively complicated formula for polynomial multiplication. We can still obtain a simple proof by proving it first for *monomials* only. Indeed, if $f = cx^k$ and $g = dx^l$, then $fg = cdx^{k+l}$, and we have,

$$(fg)(a) = cda^{k+l} = (ca^k)(da^l) = f(a)g(a).$$

Thus the assertion is true for monomials. Since any polynomial is a sum of monomials, the assertion for polynomials in general follows from what we have proved so far (both for addition and for monomials) together with the distributive property of $F[x]$. We leave the remainder of the proof for the exercises.

□

Definition 6.11 *The homomorphism ϕ of Proposition 6.15 is called the evaluation homomorphism defined by a.*

This homomorphism has some interesting properties.

Proposition 6.16 *Let F be a field, a an element of F, and let $\phi : F[x] \to F$ be the evaluation homomorphism defined by a. Then*

1. *The kernel of ϕ is the ideal $\langle x - a \rangle$ generated by $x - a$ in $F[x]$.*

2. *The quotient ring $F[x]/\langle x - a \rangle$ is isomorphic to F.*

3. *$\langle x - a \rangle$ is a maximal ideal in $F[x]$.*

Proof.

1. A polynomial f is in the kernel of ϕ if and only if $f(a) = 0$, and this is equivalent to the assertion that $x - a$ divides f (Proposition 6.14, p. 230), which is the same as saying that $f \in \langle x - a \rangle$.

2. This follows from Theorem 5.8, p. 188.

3. Since F is a field, the result follows from part 2 and Theorem 5.4, p. 179.

□

6.3.4 Reducible and irreducible polynomials

Recall that according to our terminology, we are referring to a prime polynomial in $F[x]$ as *irreducible over F*. A non-unit polynomial which is not irreducible over F is called *reducible over F*. Thus f is reducible if and only if it has a *proper* divisor, namely, one which is neither a unit nor an associate of f.

Examples

1. The reducibility of a polynomial very much depends on the field in question. Consider, for example, the second degree polynomial $x^2 + 1$ in $\mathcal{Q}[x]$. Is it reducible? Certainly not in $\mathcal{Q}[x]$ (or even in $\mathcal{R}[x]$). But it *is* reducible in $\mathcal{C}[x]$, where $\mathcal{C}$ is the field of complex numbers, since it has a root there. In fact, over $\mathcal{C}$ we have the decomposition $x^2 + 1 = (x + i)(x - i)$. What is interesting is that $\mathcal{Z}_2[x]$ also has a polynomial denoted $x^2 + 1$ and it is reducible. In which other fields is it reducible?

2. There is a similar situation with regard to rational and irrational roots. Consider the polynomial $f(x) = x^2 - 2$. It is irreducible over $\mathcal{Q}$, but reducible over $\mathcal{R}$. Can you find a smaller field, contained in $\mathcal{R}$ and containing $\mathcal{Q}$, over which this polynomial is reducible? What is the smallest such field?

3. We have seen that if a polynomial over F has a root a in F, then it is reducible over F (since $x - a$ divides it). We have also seen that the converse is not true in general, e.g., $(x^2 + 1)^2$ is reducible over R, but has no root in R. It turns out that in one rather restricted situation, reducibility is equivalent to having a root.

Proposition 6.17 *A polynomial of degree 2 or 3 over F is reducible over F if and only if it has a root in F.*

Proof. Indeed, if the polynomial is reducible, it must have a factor of degree one, hence must have a root. We leave the remaining details as an exercise.

$$\square$$

What can we say about reducibiity of polynomials and existence of roots over some particular field? For example, can we characterize all the irreducible polynomials over a given field? Since we know now that every polynomial has a unique representation as a product of irreducible polynomials, knowing all the irreducible polynomials over a given field would give us a pretty accurate picture of what *all* polynomials look like over that field. We now survey some results on reducibility and existence of roots for polynomials over the complex and over the real numbers.

The following theorem is called "the fundamental theorem of algebra", though this name reflects more the view of algebra held in past centuries than the contemporary view. It was first proved by Gauss at the age of 22, and many more proofs were found since then. We state it here without proof.

Theorem 6.5 *Every polynomial with complex coefficients and degree greater than zero, has at least one complex root.*

Remark. Because of this property, the field of complex numbers is said to be *algebraically closed*.

Corollary 6.2 *The irreducible polynomials in $C[x]$ are just the linear polynomials.*

Corollary 6.3 *Every polynomial f with complex coefficents and degree greater than 0 "splits" over C into a product of linear factors. Thus f can be written as*

$$f = c(x - z_1)(x - z_2) \cdots (x - z_n)$$

where $z_1, \ldots, z_n$ are all the complex roots of f.

Proof. Look at the decomposition of f as a product of irreducible factors. We claim that they must all be linear. For if one of the factors had degree greater than 1, it would have a root (by the previous theorem) and thus would be reducible.

$$\square$$

Next we treat polynomials over $\mathcal{R}$.

Proposition 6.18 *Let f be a polynomial over $\mathcal{R}$. If $z \in \mathcal{C}$ is a complex root of f, then $\overline{z}$ (the complex conjugate of z) is also a root of f.*

Proof. We assume that z is a root of the polynomial

$$f(x) = a_n x^n + a_{n-1} x^{n-1} + \cdots + a_1 x + a_0$$

where $a_0, a_1, \ldots, a_n \in \mathcal{R}$, and so we have

$$f(z) = a_n z^n + a_{n-1} z^{n-1} + \cdots + a_1 z + a_0 = 0$$

Conjugating both sides of the equality we have

$$\overline{a_n z^n + a_{n-1} z^{n-1} + \cdots + a_1 z + a_0} = \overline{0}$$

and making use of all the known properties of conjugates (Proposition 6.5, p. 214) we get

$$a_n \overline{z}^n + a_{n-1} \overline{z}^{n-1} + \cdots + a_1 \overline{z} + a_0 = 0$$

that is, $f(\overline{z}) = 0$. Thus $\overline{z}$ is a root of f.

$$\square$$

Theorem 6.6 *Every non-unit polynomial over $\mathcal{R}$ can be written as a product of linear factors and quadratic factors $ax^2 + bx + c$ with $b^2 - 4ac < 0$.*

Proof. Let f be any polynomial over $\mathcal{R}$. Then f is also a polynomial over $\mathcal{C}$, and hence can be decomposed over $\mathcal{C}$ as

$$f = c(x - z_1)(x - z_2) \cdots (x - z_n)$$

where $z_1, \ldots, z_n$ are all the complex roots of f and where c (being equal to the leading coefficient of f) is real.

We know from the preceding proposition that together with each complex root z of f, its complex conjugate $\overline{z}$ also appears in the above list of roots. We distinguish two cases. If the root z is real, then the polynomial $(x - z)$ is a prime factor of f also over the field $\mathcal{R}$. If, on the other hand, z is not real, then $z \neq \overline{z}$ and the quadratic polynomial $(x - z)(x - \overline{z})$ appears in the above decomposition. However, we claim that this quadratic polynomial is actually in $\mathcal{R}[x]$. Indeed, if $z = x + yi$, where $x, y \in \mathcal{R}$, then

$$(x - z)(x - \overline{z}) = x^2 - (z + \overline{z})x + z\overline{z} = x^2 - 2xz + (x^2 + y^2) \in \mathcal{R}[x].$$

Thus the prime decomposition of f over $\mathcal{R}$ consists of linear and quadratic polynomials. We also know that the quadratic factors have no real roots, hence they all satisfy the condition $b^2 - 4ac < 0$.

$$\Box$$

Corollary 6.4 *The irreducible polynomials over $\mathcal{R}$ are just the linear polynomials and the quadratic polynomials $ax^2 + bx + c$ with $b^2 - 4ac < 0$.*

Corollary 6.5 *Every polynomial of odd degree over $\mathcal{R}$ has a root in $\mathcal{R}$.*

Proof. Let f be a polynomial of odd degree over $\mathcal{R}$ and consider the decomposition of f over $\mathcal{R}$ into a product of linear and quadratic factors. The sum of the degrees of all the factors is equal to the degree of f and is therefore odd. On the other hand, the sum of the degrees of the quadratic factors is even, so that in this decomposition there must appear at least one linear factor. This linear factor has a real root, which is also a root of f.

$$\Box$$

Theorem 6.7 *Let F be a field and let $p \in F[x]$ be a polynomial over F. Then p is irreducible over F if and only if the ideal $\langle p \rangle$ generated by p is maximal in $F[x]$, in which case, $F[x]/\langle p \rangle$ is a field.*

Proof. First suppose that $\langle p \rangle$ is maximal. To show that p is irreducible we will show that every polynomial f which divides p is either a unit or an associate of p. Indeed, if $f|p$ then by Proposition 6.10, p. 219,

$$\langle p \rangle \subseteq \langle f \rangle \subseteq F[x].$$

Since $\langle p \rangle$ is maximal, it follows that $\langle f \rangle$ must be equal to either $\langle p \rangle$ or $F[x]$. Accordingly, f is either an associate of p or a unit.

(Note that this part of the proof makes no use of any special properties of the polynomial ring, and therefore holds in any integral domain.)

Conversely, assume that p is irreducible, and let I be an ideal in $F[x]$ such that

$$\langle p \rangle \subseteq I \subseteq F[x].$$

Now, by Theorem 6.1, p. 223, I is principal — that is, we have $f \in F[x]$ with $I = \langle f \rangle$. From $\langle p \rangle \subseteq \langle f \rangle$ it follows that f divides p so that f must either be a unit or an associate of p. Accordingly, we have that either $I = F[x]$ or that $I = \langle p \rangle$. Thus $\langle p \rangle$ is maximal.

The last statement follows from Theorem 5.4, p. 179.

$\square$

Example. We can use this result to construct a field with 8 elements. Begin with $\mathcal{Z}_2$ and consider the polynomial $p \in \mathcal{Z}_2[x]$ given by $p(x) = x^3 + x^2 + 1$. Since the two elements of $\mathcal{Z}_2$, 0 and 1, are not roots of p, it follows from Remark 2, p. 231 that p is irreducible. Hence, $E = \mathcal{Z}_2[x]/\langle p \rangle$ is a field. How many elements does it have? We will show that the cosets of the 8 polynomials of degree less than or equal to 2 are distinct elements of E and they give all of E.

If $f, g \in \mathcal{Z}_2[x]$ and both have degree not more than 2, then so does their difference and hence it cannot be divisible by p. Therefore, the 8 cosets are distinct. Now take any $f + \langle p \rangle \in E$. If the degree of f is more than 2, it can be divided by p to obtain a remainder g which is a polynomial of degree less than or equal to 2, that is, one of our 8 polynomials. But, $f + \langle p \rangle = g + \langle p \rangle$ and so we have all the cosets.

Could we use this method to get a field with 9 elements? How about 10 elements? You will have a chance in the exercises to muse about just how general is the situation represented by this example.

6.3.5   Extension fields

If you have a polynomial over a field, does it have at least one root? You can already start to analyze this question using your previous mathematical knowledge. As you know, the polynomial $x^2 + 1$ does not have a root in the field $\mathcal{R}$ of real numbers. But you also know that this is not the whole story. Suppose you considered the field $\mathcal{C}$ of complex numbers. This contains $\mathcal{R}$ as a subfield and so we can view $x^2 + 1$ also as a polynomial in $\mathcal{C}[x]$. Now the polynomial $x^2 + 1$ *does* have a root in $\mathcal{C}$, namely $i = \sqrt{-1}$.

This raises two questions. First, what is really going on mathematically when you "extend the real numbers to the complex numbers so as to find a solution to $x^2 + 1 = 0$"? And second, how general is this situation; that is, when a polynomial fails to have a root in a certain field, can we always "extend" this field to a larger field in which the given polynomial does have a root? We will try to answer both questions with a single analysis.

Definition 6.12 (extension field.) *A field E is said to be an* extension field *of a field F if F is a subfield of E.*

The next theorem, sometimes called the *Fundamental Theorem of Field Theory* gives an abstract version of finding a solution to $x^2 + 1 = 0$ by extending the real numbers to the complex numbers. The interesting point is that this process works for *any* (non-constant) polynomial over *any* field. The intuitive idea behind the construction is that in order to "adjoin" to F a root of an irreducible polynomial p, we first adjoin an "indeterminate" x obtaining the polynomial ring $F[x]$) and then "identify" $p(x)$ to 0 by taking the quotient ring $F[x]/\langle p \rangle$. Since, under the above identification, the image of $p(x)$ is 0, the image of x must be a root of p.

Theorem 6.8 *Let F be a field and $f \in F[x]$ a polynomial of degree at least 1. Then there is an extension field E of F such that f has a root in E.*

Proof. Let F be a field and $f \in F[x]$ a polynomial of degree at least 1. Let p be a an irreducible polynomial that divides f. Then it suffices to show that there is an extension field in which p has a root.

It follows from Theorem 6.7 above that the quotient $F[x]/I$, where $I = \langle p \rangle$, is a field. Let us call it E. We will show that E contains a subfield which is isomorphic to F. Then by identifying F with this subfield, we may view E as an extension field of F. We start with the canonical homomorphism $\tau : F[x] \to E$, defined as usual (see Section 5.3.5, p. 188) by $\tau(g) = g + I$. We will now show that the restriction of τ to F (which we still call τ) is the required isomorphism. That is, we let $\overline{F} = \{a + I : a \in F\}$, and show that the map $\tau : F \to \overline{F}$ defined by $\tau(a) = a + I$ is an isomorphism. Because τ is the restriction of the canonical homomorphism to F it is clearly a homomorphism. It is also clearly *onto*. It remains to show that τ is one-to-one. Indeed, if $\tau(a) = \tau(b)$ then $a + I = b + I$, and so $a - b \in I$, which means that $p|a - b$. This implies that $a - b = 0$, for otherwise we would have a polynomial of degree at least 1 (p) dividing a polynomial of degree 0 $(a - b)$, which is a contradiction. Thus $a = b$ and τ is one-to-one.

Because of the isomorphism, we will henceforth take the liberty of identifying the elements $\overline{a} = \tau(a) \in \overline{F}$ with the corresponding elements $a \in F$. In particular, this allows us to view p as a polynomial over E, by replacing each coefficient a with its image $\overline{a}$ under τ.

We will now show that p has a root in E; specifically, we will show that $\overline{x} = x + I \in E$ is a root of p. Since x refers here to a specific element in the ring $F[x]$, and not to a general "variable", we will use y for the variable to avoid confusion. Thus we want to show that if we substitute in the polynomial p the element $\overline{x} = x + I \in E$, then we get 0. To this end we write out the polynomial p,

$$p(y) = a_n y^n + a_{n-1} y^{n-1} + \cdots + a_1 y + a_0$$

and substitute $\overline{x} = \tau(x)$ in it. Using the identification $a = \tau(a)$ and the fact that τ is a homomorphism we get:

$$
\begin{aligned}
p(\tau(x)) &= a_n \tau(x)^n + a_{n-1} \tau(x)^{n-1} + \cdots + a_1 \tau(x) + a_0 \\
&= \tau\left(a_n x^n + a_{n-1} x^{n-1} + \cdots + a_1 x + a_0\right) \\
&= \tau(p) = p + I,
\end{aligned}
$$

which is the zero element in E since $p \in I$. This completes the proof.

$\Box$

Theorem 6.9 *Let F be a field, $p \in F[x]$ an irreducible polynomial of degree $n \geq 1$. Let $E = F[x]/\langle p \rangle$ and $\alpha = x + \langle p \rangle$. Then every element of E has a unique representation as a polynomial in α of degree less than n with coefficients in F.*

Proof. We know from the previous theorem that E is an extension field of F, and that α is a root of p in E.

We let, as before, τ be the canonical homomorphism from $F[x]$ onto E. Then for every $f \in F[x]$, $\tau(f) = f + \langle p \rangle$. In particular we have $\tau(x) = \alpha$ and $\tau(a) = a$ for all $a \in F$ (recall from the proof of Theorem 6.8 that we identify $\tau(a)$ with a). If we now take any element of E, it can be expressed in the form $\tau(f)$ for some $f \in F$. We do a division-with-remainder of f by p getting $f = qp + r$ for suitable polynomial $q, r \in F[x]$, such that either $r = 0$ or $\deg(r) < n$. Now $f - r = qp \in \langle p \rangle$, hence $\tau(f) = \tau(r)$. Thus every non-zero element of E has the form $\tau(r)$ where r is a polynomial of degree less than n. But τ is a homomorphism, so if we let

$$r = a_t x^t + \cdots + a_1 x + a_0$$

then we have

$$\tau(r) = \tau(a_t)\tau(x^t) + \cdots + \tau(a_1)\tau(x) + \tau(a_0) = a_t \alpha^t + \cdots + a_1 \alpha + a_0$$

which is of the required form. To show uniqueness, assume that we have two polynomials $f, g \in F[x]$ of degree less than n, and assume that $f(\alpha) = g(\alpha)$. This means $\tau(f) = \tau(g)$, hence p divides $f - g$. Since p cannot divide a polynomial of degree less than n, it follows that $f - g = 0$, that is, $f = g$.

$\square$

Construction of the complex numbers

As an application, let's consider the case where $F = \mathcal{R}$, the field of real numbers, and $p = x^2 + 1$, which is an irreducible polynomial. We let $\mathcal{C} = \mathcal{R}[x]/\langle x^2 + 1 \rangle$, and we let $i = \tau(x)$ (same as α above). Then we know that $\mathcal{C}$ is a field and $i \in \mathcal{C}$ is a root of the polynomial $x^2 + 1$ (i.e., a square root of -1). Also, by Theorem 6.9, every element of $\mathcal{C}$ has a unique representation as a polynomial in i of degree smaller than 2, in the form $a + bi$. Thus we have constructed the field of complex numbers. You are invited to write down the formulas for addition and multiplication in $\mathcal{C}$ (they are the same as in $\mathcal{R}[x]$, except that the result is taken "mod p"). You'll find that they are the same as the familiar operations on the complex numbers.

6.3.6 Splitting fields

In the previous section it was established that a given polynomial f over a field F will always have a root α in an appropriate extension field E.

We then know that f will factor over E in the form $f = (x - \alpha)g$. Is it possible to find an extension field in which f factors all the way down to linear factors? In how many different ways can this be done? How does all of this very abstract stuff look in specific cases? We will look briefly at these questions in this section.

Definition 6.13 *A polynomial of degree 1 is said to be* linear. *A polynomial is said to* split *over a field if it can be written as a product of linear polynomials over that field.*

Let F be a field and $f \in F[x]$. An extension field E of F is said to be a splitting field *for f over F if f splits over E and f does not split over any proper subfield of E.*

Let F be a field. Two extension fields of F, E_1 and E_2, are said to be isomorphic over F *if there exists an isomorphism $\phi : E_1 \to E_2$ such that $\phi(a) = a$ for every $a \in F$. Such a map ϕ is called an* isomorphism over F.

Remark. Notice that because the degree of the product of polynomials over a field is the sum of the degrees of the individual polynomials, then the number of linear factors (counting each factor according to the number of times it occurs) in a polynomial that has split is equal to the degree of the polynomial.

Theorem 6.10 *Let F be a field. Then every polynomial over F of degree at least 1 has a splitting field over F. The splitting field is unique up to an isomorphism over F.*

Proof. Given a non-constant polynomial f over F, it follows from repeated application of Theorem 6.8 that there is an extension field E_1 of F in which the polynomial factors into linear factors; for as long as it doesn't, we can apply the theorem again to adjoin another root. Thus we have the following decomposition over E_1:

$$f = a(x - \alpha_1)(x - \alpha_2) \cdots (x - \alpha_n),$$

where the α_is are the roots of f in E_1, and may be appearing in this decomposition more than once.

To find a *minimal* such field, we use a standard device. We take E to be the intersection of all the subfields of E_1 which contain F and also contain all the αs. It is then straightforward to check that E is a splitting field of f over F (Exercise 23).

The proof of the uniqueness assertion is somewhat subtle and will be omitted.

$$\square$$

Example. What happens if this proposition is applied to the field $\mathcal{R}$ and the polynomial $x^2 + 1$? Do you see that the splitting field we get is just the complex numbers? Thus, this theory gives us a very general machine, for which constructing the complex numbers is just a special case.

6.3.7 EXERCISES

1. Prove part 3 of Proposition 6.14, p. 230 using mathematical induction and without using the prime decomposition property.

2. Complete the proof of Proposition 6.15, p. 232.

3. Complete the proof of Propsition 6.17, p. 234.

4. Show that $Z[x]$ is not a Euclidean domain. (Hint: One way of showing this is by using degree arguments to show that the GCD of 2 and x, which is 1, cannot be written as $a \cdot 2 + b \cdot x$, where a and b are integers.)

5. Show that the polynomial $x^3 + x^2 + 1$ is irreducible in $Z_2[x]$.

6. Find as many fields as you can over which the polynomial denoted $x^2 + 1$ is reducible.

7. Try to find a field with 9 elements.

8. Try to find a field with 10 elements.

9. Work out the details of using splitting fields to construct the complex numbers.

10. Show that $x - a$, where $a \in F$, is always prime in $F[x]$.

11. Use the tools of Section 6.2 to build a multiplication table for $Z_3[x]$. Find all irreducible polynomials of given degree d for as high a value of d as you can go.

12. Show that, for every prime p there exists a field of order p^2.

13. Find all the zeros of $x^5 + 4x^4 + 4x^3 - x^2 - 4x + 1$ over Z_5.

14. Imitate the construction of Z_p and the proof that it is a field, to obtain an elementary proof of Theorem 6.6, p. 235. That is, construct the ring $F[x]_p$ of "remainders" of the polynomial ring $F[x]$ modulo the prime polynomial p, and prove that it is a field extending F and that it has a root for the polynomial p. Following is a more detailed description of what you have to do in this exercise.

    (a) Define $F[x]_p$ as the set of all polynomials in $F[x]$ whose degree is smaller than $\deg(p)$. Define in $F[x]_p$ addition $+_p$ and multiplication $\cdot_p$ by performing the usual addition and multiplication in $F[x]$, dividing the result by the polynomial p and taking the remainder.

    (b) Show that $F[x]_p$ is a homomorphic image of $F[x]$ and deduce that it is a commutative ring with 1.

(c) Show that $F[x]_p$ is a field. (This is the first time you'll need to use the assumption that p is irreducible.)

(d) Show that F can be embedded in $F[x]_p$ and that, under this embedding, x (as an element of $F[x]_p$) is a root of p.

15. Show that $Z[i]/\langle 2 + i \rangle$ is a field. How many elements does it have?

16. Imitate the construction of the complex numbers on p. 239 to show that $Q[x]/\langle x^2 - 2 \rangle$ is a field containing Q and a root of the polynomial $x^2 - 2$. Show further that this field is isomorphic to the subfield $\{a + b\sqrt{2} : a, b \in Q\}$ of $\mathcal{R}$.

17. Find the splitting field of $x^4 + 1$ over Q.

18. Find the splitting field of $x^4 + x^2 + 1$ over Q.

19. Let F be a field and let $f \in F[x]$ be irreducible over F. Suppose that a and b are roots of f in some extension field E of F. Let F_a be the smallest subfield of E containing a and similarly for b. Show that there is an isomorphism from F_a onto F_b that is the identity on F and maps a to b.

20. Is the isomorphism in the previous exercise unique?

21. Let F be a field and let $f \in F[x]$. If g is the polynomial obtained by substituting $x + 1$ for x in f, does g have the same splitting field over F as does f?

22. Let F be a field. Two polynomials $f, g \in F[x]$ are said to be *relatively prime* if there is no polynomial of positive degree in $F[x]$ that divides both of them. Show that this property is unchanged on passing to an extension field of F.

23. Show that the field E defined in the proof of Theorem 6.10, p 240, is a splitting field of f over F.

Index

Elizabeth Lawrence

The Little Bulbs

A Tale of Two Gardens

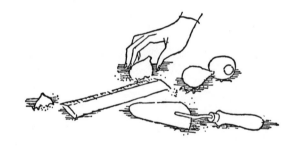

Introduction by **Allen Lacy**

Duke University Press   Durham 1986

To
Carl
With Love

Contents

Introduction

Books on gardens and gardening, like all other books, have their seasons. They come into print and they go out of print. Sometimes they march to the remainder table with frightening speed, and judging from the great numbers of review copies of garden books I receive each year from American trade publishers who hope for some mention of their current crop in the column I write for *The Wall Street Journal*, a swift demise is in certain cases entirely condign. The unfortunate fact, however, is that really good books about gardening also go out of print—little treasures that we hear about too late. But with really good books about gardening, out of print does not mean out of date, which explains the tendency of dedicated horticultural readers to duck into every secondhand bookstore they come across, in hope of discovering a copy of Alice Morse Earle's *Old Time Gardens* (1901), Buckner Hollingsworth's *Flower Chronicles* (1958), or Beverley Nichols' *Down the Garden Path* (1932), to name just three of my own triumphant finds.

But on some rare and fine occasions, a really good book about gardening will get a second chance and a new circle of readers to delight. Such is the case with the late Elizabeth Lawrence's *The Little Bulbs: A Tale of Two Gardens*, originally published in 1957 by Criterion Books and now reissued by Duke University Press.

As anyone who has ever read six consecutive sentences by Elizabeth Lawrence will have realized instantaneously, she is a "garden writer," only in the way that M. F. K. Fisher is a "food writer." Miss Lawrence transcends the usual category, for precisely the reason she states in her original preface to *The Little Bulbs*: "Gardening, reading about gardening, and writing about gardening are all one; no one can garden alone."

No one can garden alone—or put another way, gardening is a powerful means of escaping from the alienation that social critics and philosophers since Soren Kierkegaard have described as epidemic in the modern age. To garden means to be connected—connected to the earth, to all that grows here, and to the seasons and the years. To garden means to be indebted. Some of our debts we know and can acknowledge. We can thank the long line of plant explorers that includes Ernest ("Chinese") Wilson, who brought us Regal lilies and other fine ornamentals from the Asian mainland, and David Fairchild, who brought us Japanese flowering cherries (even if he also brought us the aggressive and invasive kudzu vine that now festoons much of the South, strangling less luxuriant forms of vegetation). We can thank the twentieth-century hybridizers who have gilded the daylily and improved the peony, even if they have also stolen fragrance from the rose. As individuals, we can express our gratitude to the next door neighbors who gave us our start of an especially fine perennial aster or to the gardener-friend several states away who sent us a collection of sedums and sempervivums.

In *The Little Bulbs*, Elizabeth Lawrence focused on two things. First, there were the little bulbs themselves—not only common things like aconites and crocuses and the tinier sorts of daffodils, but also rarities like *Clinopodium georgianum* and *Habranthus texanus*—all plants which she grew (but sometimes lost) in her garden, first in Raleigh and then in Charlotte. Second, and of equal importance, there were the friends who shared her passion for gardening—Miss Caroline Dormon of Louisiana, Mr. William Lanier Hunt of North Carolina, Miss Willie Mae Kell of Texas, and especially Mr. Krippendorf of Ohio, whose garden in Cincinnati is the second of the two gardens referred to in the subtitle of *The Little Bulbs*.

Mr. Krippendorf is a faintly mysterious figure in this book, first because his first name is never used, second because he was obviously able to plant by the thousands and tens of thousands bulbs that most of us can only afford to plant by the dozen, if that many. Miss Lawrence most decorously did not disclose the size or the source of the fortune that provided him with a garden taking in hundreds of acres of mature woodland, for it was his plants and his interest in plants that made them friends and epistolary garden companions.

Thanks to *The Little Bulbs*, Mr. Krippendorf became something of a legendary figure among its readers. Some people even thought he might be mythical or fictional, like crusty old Amos Pettingill, who writes the catalog for White Flower Farm in Litchfield, Connecticut, but doesn't really exist. In *Lob's Wood*, a further account of her friend and the garden he made, written after his death and published in 1971 by the Cincinnati Nature Center, Elizabeth Lawrence wrote:

> Wherever I go, people ask me about Mr. Krippendorf. Some want to know whether I made him up (as if I could!). Once when I was lecturing at the University of Mississippi, not long before he died, I was told that the college engineer wanted to meet me. When he was introduced I said, "What kind of garden have you?" He said he didn't have any. "Then how did you happen to come to the lecture?" "I came because I have something to ask you," he said. "How is Mr. Krippendorf?" When I told him that he was not at all well, he looked troubled. "I was afraid of that," he said.

Time is not always kind to the gardens that passionate

gardeners bring into being and lovingly tend. I've known a few people who ended their lives on this earth anxious and fretful that their gardens would not long survive them. It is good, therefore, to report that as the Cincinnati Nature Center Mr. Krippendorf's wooded garden lives on, offering visitors its peaceful shaded paths in summer and its sheets of aconites and daffodils and cyclamens during their seasons of bloom. It also survives in the pages of *The Little Bulbs*, together with Miss Lawrence's own, much smaller gardens in the North Carolina piedmont.

One final, editorial note. The list of sources for bulbs at the end of the book has been revised. Nurseries, like gardeners and gardens, have their times and seasons. Sometimes they go out of business, and often they are not replaced with their like. Some of the bulbs that are mentioned in this book are easily obtained. Some, if they are found at all, may turn up only after protracted search. But looking for plants, sometimes without success, is a simple fact of every true gardener's life. We may still be very grateful to Elizabeth Lawrence for writing so winningly about the little bulbs that offer such enormous pleasure in disproportion to their size.

Allen Lacy
Professor of Philosophy
Stockton State College
Pomona, New Jersey

Preface

This is a tale of two gardens: mine and Mr. Krippen-dorf's. Mine is a small city back yard laid out in flower beds and gravel walks, with a scrap of pine woods in the background; Mr. Krippendorf's is hundreds of acres of virgin forest. Both are perfect for little bulbs, for no garden is too small to hold them all if only a few of each are used, and no forest is too large to show them off if enough of one kind is planted. While I content myself with a single daffodil, Mr. Krippendorf likes to see ten thousand at a glance. Where I plant half a dozen, he plants a hundred. "I had another lovely day," he wrote once in November; "it is perfect Indian summer weather, so I got out fairly early (7 A. M.) and planted more than two hundred and fifty small bulbs in the open." Every fall for fifty years he has planted bulbs under his beech trees. And the plantings have multiplied.

I came to know Mr. Krippendorf when he saw an

article that I had written for *Herbertia* * on amaryllids in North Carolina, and he wrote to tell me about amaryllids in Ohio. Now, for more than a decade, we have compared flowers and seasons. Rereading his letters I find recorded the beauty of ten Ohio springs—and ten summers and falls and winters as well, for there is never a time when he cannot find something of beauty or interest in his woods, even when there is no bloom. "I do think the last two days were the darkest I ever saw," he wrote once in February. "The ground leaves have dried out, and have not the rufous color they have when wet. When wet, they seem to emanate light."

Sometimes I am not so sure which is more real to me: finding an early flower in my own garden, or following Mr. Krippendorf's solitary ramblings across his wooded hills. At the top of one hill there is the house, and at the bottom is a clear, wide creek. On the far side of the creek the limestone banks are hung with ferns and wild flowers. On the near side is a little green meadow, long and narrow and embroidered with blue phlox. A broad path winds from the house to the meadow, with gray bridges across the ravines. In the steep places are steps made of great, flat stones drawn up from the creek bottom. From the main path, tributaries lead to other parts of the woods. Along these you can walk up and down hill for hours and never come to an end of squills and daffodils. When the leaves are off the trees, you can see through the branches to other hills, far away. Those

* *Yearbook of the American Amaryllis Society.*

2

hills are wooded, too, and I suppose they are full of bulbs. There is nothing else in sight—only Mr. Krippendorf in his leaf-colored jacket, with a red bandanna about his neck.

There are other gardens and gardeners that must enter into this tale. Some I have seen, and some I know only from letters or books, but all of these gardens are as real to me as my own. When a little bulb blooms in my garden, I see it not only there, but also beside a Florida lake, or on a rock wall in Alabama, or in a pot on a San Francisco terrace. A celestial reminds me of the one so lovingly painted by Miss Caroline Dormon for her *Wild Flowers of Louisiana,* and alliums make me think of Miss Willie Mae Kell's little garden in northern Texas. Miss Kell says her garden is the size of a pocket handkerchief, but judging by all that comes out of it, it must have more plants per square inch than any other spot in the world.

A crocus that blooms every spring recalls hours spent in Elizabeth Rawlinson's rock garden in the Shenandoah Valley, where so many rare bulbs were crowded in with endemics transplanted from the nearby shale barrens. Cyclamens make me think of afternoon tea with Mrs. De Bevoise, and of the flowering walls at Cronamere, her garden on Long Island Sound. These two gardens are of the past, but they still bloom in my mind, and must be quick in the memory of any other gardeners who knew their generous owners.

Just as vividly before me are the gardens of books.

"The Secret Garden," that sweet, mysterious place enclosed by high walls on which ivy and running roses were matted together; "Elizabeth's" cold acres in northern Germany; and Mrs. Loudon's tidy borders in Bayswater, London. A little more than a hundred years ago Mrs. Loudon was one of the "very few indeed" who grew bulbs out of doors. She planted the half-hardy as well as the hardy, and left them to brave the rigors of a London winter. She grew more bulbs, wrote more interesting things about them, and gave more practical and explicit advice about growing them than any other writer I know. I think I refer to *The Ladies' Flower Garden of Ornamental and Bulbous Plants* (London, 1841) as often as any other bulb book—except perhaps Mrs. Wilder's *Adventures with Hardy Bulbs.*

I never knew Mrs. Wilder myself, but when I asked Mr. Krippendorf if he were the "friend from Ohio" that she mentions from time to time, he replied that he probably was. "I used to see her quite often in New York," he wrote. "She would have dinner with me, and then we would sit in the Saint Regis and talk about plants, and I would take her to her train for Bronxville at nine o'clock. Her garden was small but full of good things."

It is not enough to grow plants; really to know them one must get to know how they grow elsewhere. To learn this it is necessary to create a correspondence with other gardeners, and to cultivate it as diligently as the garden itself. From putting together the experiences of gardeners in different places, a conception of plants

Preface

begins to form. Gardening, reading about gardening, and writing about gardening are all one; no one can garden alone.

Elizabeth Lawrence
Charlotte, North Carolina.

1

The First Flower

And since to look at things in bloom
Fifty springs are little room,
About the woodlands I will go
To see the cherry hung with snow.
　　　A. E. Housman, *A Shropshire Lad*

As soon as spring is in the air Mr. Krippendorf and I
begin an antiphonal chorus, like two frogs in neighbor-
ing ponds: What have you in bloom, I ask, and he an-

swers from Ohio that there are hellebores in the woods, and crocuses and snowdrops and winter aconite. Then I tell him that in North Carolina the early daffodils are out but that the aconites are gone and the crocuses past their best.

"You complain of not having color in your garden," he once wrote on the last day of January. " 'I have known greater wrongs, I that speak to you' (but you wouldn't know that, as I am quite sure you were not brought up on McGuffey's readers). Here, everything is still encased in ice from last Saturday's freezing rain. We had a snowstorm in the night, and are promised more sleet tomorrow. As for color in the woods, nothing shows above the dead leaves."

But even in Ohio there is bloom sometimes before January is out, and another year on the twenty-sixth of January he wrote: "I hope spring is waking up for you. With us it is just turning over in its sleep. The weather has been warm (up to fifty) and one can find an occasional snowdrop in the woods, or see a bit of yellow where the aconite should be. These small early flowers give me a strange pleasure that many of the later and handsomer ones do not, especially those among them which have self-sown and which I find in unexpected places. Sunday I saw a small bulb heaved out of the ground, and when I picked it up I found a lavender tip between the leaves, so it won't be long now."

By the fifteenth of February he could report winter aconite, crocuses, *Colchicum luteum*, hellebore, and the broad-leaved snowdrop in bloom, and the narrow-leaved snowdrops showing thousands of little white points. On the twenty-fourth he wrote: "I have your letter—we, too, have lots of bloom, mostly bulbous. The winter aconite is nearly gone, but the snowdrops are just in their prime; also snowflakes, *Scilla sibirica*, *S. bifolia* and Spring Beauty, white grape hyacinths, *Hyacinthus azureus*, many *Iris reticulata*, and more crocuses than usual." Another year on the same date he wrote, "Now the winter flowers are opening, and today I saw two yellow colchicums which I imported from India—why I should have them after fifteen or twenty years I do not know." This interested me, too, for the species comes from elevations of seven thousand feet in the Himalayas and is considered difficult in gardens. However, Billy Hunt says he has now had it for several years in Chapel Hill, where it blooms in January.

Sometimes when the late winter is severe Mr. Krippendorf has more flowers than I. In one of these seasons he wrote on the tenth of February: "I was surprised to hear of the paucity of bloom in your garden, as I once read a book by an Elizabeth Lawrence who listed quantities of plants that bloomed in February or even January in her garden (which she alleged was in Raleigh, North Carolina). We have quite a few snowdrops now, and some eranthis, in spite of the fact that the pool on the terrace freezes every night." And later: "I have your letter dated Fourth Sunday in Lent but not mailed until Tuesday. You say you might as well have lived in Ohio this winter—that sounds almost scornful. Yesterday was a wonderful day, not too warm, and sunshine off and on. I have tens of thousands of winter aconites in the woods—bold groups repeating themselves into the distance and small clumps appearing almost anywhere. There were lots of snowdrops, also the spring snowflakes, and *Adonis amurensis.*"

Mr. Krippendorf says that *A. amurensis* is his earliest flowering herbaceous plant. He has reported it as early as the seventh of February. One year he wrote on the seventeenth: "This has been a wonderful day of winter sunshine, and I have found a number of things in bloom. In the morning I ran across a perfect flower of adonis—the semi-double form, not the monstrous double one from Japan which comes a month later. The single form is one of the loveliest flowers I know. There are two survivors of my original planting of forty years ago."

The single-flowered adonis is my favorite, too, but the blossoms of the double ones last longer. I have known them to last nearly a month. The metallic luster of the green-gold petals seems to protect them from the weather, and if the frost lays them flat on the ground they spring up again as soon as it turns warm. With me, both forms bloom in February, but once I found a bright flower on the twenty-second of January. The finely cut, feathery foliage improves with the years, making a handsome clump of greenery at the time when green is most welcome. It lasts only a few weeks, but by the time it is gone there are plenty of other green things growing. Although it lives long when once established, I have not found the adonis an easy plant to keep, and once it is settled it does not like to be disturbed. For some years now it has flourished in a low part of the garden where there is a good deal of shade from the high branches of old pine trees and a good mulch of their needles to keep the ground moist in summer. Here they bloom regularly beside a drift of scilla Spring Beauty.

Two blue-flowered perennials sometimes bloom in my garden in February. *Anchusa myosotidiflora* is one. Its pale forget-me-not flowers are pleasing with *Hyacinthus azureus* and the indispensable early spring annual called baby-blue-eyes (nemophila)—but are far from pleasing with the blue-violet of scillas and anemones. Its flowers appear for two months or more, and after that the large round leaves are a strong summer accent. The flowers of *Pulmonaria angustifolia* are almost as blue, but a deeper tone. They sometimes appear in my garden in mid-February, and there is a variety *azurea* (if it can be found) that is said to be earlier and bluer. The dark green leaves make a close mat which increases slowly and steadily. Both of these are good with the early bulbs because they like shade and leafmold.

In a February letter Mr. Krippendorf says: "You belittle hellebores in spring, but if you saw a Lenten rose in bloom just four days after nine weeks of snow I think you would appreciate it. I cut twenty-four flowers from one plant, and did not take them all."

I *do* appreciate hellebores in spring, now that I know that *Helleborus orientalis* does not wait for Lent, even when Lent comes early, but blooms in North Carolina soon after the new year. However, it was a long time before I had bloom in January, for the seedlings usually flower later until they are well established. My first seedlings came from Mrs. Hassan's garden in Alabama. "Their flowers vary in color from greenish-white and brown and pink and lavender to the darkest shades of plum and burgundy," she wrote. "The pink one is a star performer, almost four inches across, speckled with dark red spots and washed with apple green."

Since the Christmas rose (*Helleborus niger*) has never flourished with me, I decided that it does not like the South, but I now find it grows well and blooms freely in other gardens in Charlotte, and I am told that lime is what it needs. The time of bloom depends upon the age of the plant, the situation, and also the form. *H. niger praecox* is the early form that blooms before Christmas.

February also finds one of the best of the little bulbs in bloom, *Anemone blanda,* the Grecian windflower. It flowers so long and is so easily grown that it is hard to understand why it is so little planted, especially as the tubers are cheap enough to allow planting in quantity. It is well to begin with a lot, for they will be wanted once they are seen. In suitable places, established clumps seed themselves, but this takes time. The tubers should be planted two inches deep in good leafy soil, where they will get enough sun to open the flowers in winter—for they open only in the sun—but not so much that it dries them out in the summer. The many-petalled flowers are like enormous hepaticas, and in the variety *atrocaerulea* they are as blue as gentians. There is also a rose-colored form—or so they say—and a series of pinks, lilacs, and pale blues. One year they bloomed from the last of January to the second week in April. The prettily cut, three-lobed leaves come with the flowers and last until hot weather dries them up.

A. *apennina* begins to bloom just as the Grecian windflower is tapering off. It is much the same, except that the flowers are smaller, and its chief value is to prolong the season. A. *nemorosa,* the European wood anemone, blooms about the same time. This one likes shade.

Ranunculus ficaria, the lesser celandine, has been singled out by both Mr. Krippendorf and Mr. Wordsworth for its early brightness. Mr. Krippendorf wrote from Ohio: "A little European weed, *Ficaria,* has been increasing on a hillside for many years. It makes a dense mat of brightest green studded with yellow spring buttercups; the whole thing only an inch high. The way it drifts down the slope is most intriguing." And Mr. Wordsworth wrote from Grasmere (though not to me): "It is remarkable that this flower, coming so early in the spring as it does, and so bright and beau-

tiful, and in such profusion, should not have been noticed earlier in English verse." Some say that it was
noticed earlier, and by a better poet than Mr. Wordsworth.

> *Cuckoo-buds of yellow hue*
> *Do paint the meadows with delight.*

Mr. Wordsworth makes up for early neglect by writing a long poem "To the Small Celandine," and two
more "To the Same Flower." "What adds much to the
interest that attends it," he says in introduction, "is its
habit of shutting itself up, and opening out according
to the degree of light and temperature of the air."

> *Thou dost play at hide-and-seek;*
> *While the patient primrose sits*
> *Like a beggar in the cold. . . .*

He notes its habitat:

> *Careless of thy neighborhood*
> *Thou dost show thy pleasant face*
> *On the moor, and in the wood,*
> *In the lane. . . .*

The lesser celandine blooms

> *As soon as gentle breezes bring*
> *News of winter's vanishing*

The First Flower

. . . which in Grasmere, according to the poet, is in the month of February.

Once I found the first flower on Washington's Birthday in my own garden, but usually I must wait until the first of March for the small round bud to open. Then I know that in a very short time Mr. Krippendorf will be coming home from a walk, with a flower in his buttonhole.

In Elizabeth's cold German garden, near the Baltic Sea, *R. ficaria* bloomed in May: "The dandelions carpeted the three lawns—they used to be lawns, but have long since blossomed out into meadows filled with every sort of pretty weed—and under and among the groups of leafless oaks and beeches were blue hepaticas, white anemones, violets, and celandines in sheets. The celandines particularly delighted me with their clean, happy brightness, so beautifully trim and newly varnished, as though they, too, had had the painters at work on them. Then, when the anemones went, came a few stray periwinkles and Solomon's Seal, and all the bird cherries blossomed in a burst."

Mrs. Wilder was another lover of this small golden weed: "One plant in particular I have found especially happy as a ground cover for *Scilla sibirica*. The smaller celandine . . . is quite ideal, preferring to grow, as do the scillas, in partial shade. It forms a flat, close, advancing mat of small shining leaves spattered over (while the scillas are in bloom) with brilliantly yellow, highly varnished stars."

15

It might be well to bear in mind the hint about the "flat, close, advancing mat." It is best to keep the celandine out of the rock garden and away from small and rare bulbs, restricting it to areas where an indestructible ground cover is in demand. I once measured a small clump after it had been set out for a year, and it had a diameter of eighteen inches. As a winter ground cover it has virtue, for the round leaves come up in December and stay green through all kinds of weather, looking so crisp that I once thought of putting them in a salad. One nibble put an end to *that* thought. If, as a child, you were ever dared to eat a leaf of that dear Victorian plant called elephant ears, and if you were foolish enough to take the dare, you will know why I decided (like the rabbits) to let it be.

In spite of its prominence in literature, the lesser celandine is still unappreciated by gardeners, and I can find no commercial source for it. But, as Mr. Wordsworth says:

> *Praise enough it is for me*
> *If there be but three or four*
> *Who will love my little flower.*

"If you don't like yellow, you won't like the woods now," Mr. Krippendorf wrote in the middle of March. "There are tens of thousands of winter aconite in bloom, so thick that one can see them a hundred yards away. I love to see the patches of color repeat themselves in the